CHINA'S DISAPPEARING COUNTRYSIDE

T0250898

China's Disappearing Countryside
Towards Sustainable Land Governance for the Poor

YONGJUN ZHAO
University of Groningen, the Netherlands

Routledge
Taylor & Francis Group

LONDON AND NEW YORK

First published 2013 by Ashgate Publishing

Published 2016 by Routledge
2 Park Square, Milton Park, Abingdon, Oxfordshire OX14 4RN
711 Third Avenue, New York, NY 10017, USA

First issued in paperback 2016

Routledge is an imprint of the Taylor & Francis Group, an informa business

British Library Cataloguing in Publication Data
Zhao, Yongjun.
  China's disappearing countryside : towards sustainable land
  governance for the poor.
  1. Land tenure--China. 2. Land reform--China. 3. Rural
  development--China. 4. Agriculture and state--China.
  5. Rural poor--China.
  I. Title
  333.3'151'086942-dc23

The Library of Congress has cataloged the printed edition as follows:
Zhao, Yongjun.
  China's disappearing countryside : towards sustainable land governance for the poor / by
Yongjun Zhao.
    p. cm.
  Includes bibliographical references and index.
  ISBN 978-1-4094-2821-3 (hardback) 1. Land reform--China. 2. Rural poor--China.
  3. Sustainable development--China. I. Title.
  HD923.Z4744 2012
  333.3'151--dc23
                                                                          2012026014

ISBN 13: 978-1-138-27723-6 (pbk)
ISBN 13: 978-1-4094-2821-3 (hbk)

Contents

List of Figures and Tables

Figures

Tables

Preface and Acknowledgements

Despite its unprecedented achievements in rural development, China remains a lower-middle income country. China is facing mounting challenges of equitable and sustainable development whereby land tenure continues to be a perplexing issue yet to be effectively tackled. The trajectory of China's land reform is a contradiction in terms – the reform through both collectivization and ongoing decollectivization has not proven effective in safeguarding the interests of the poor. Unsound practices in farmland use and management have contributed to rising farmland loss, social conflicts and deprivation of the landless, which perpetuate rural poverty.

The current hybrid land tenure systems characterised by collective ownership and individual use rights exert both positive and negative effects on land governance. China's approach to land laws, policies and institutional reforms is characterised by its inherent weaknesses which impede the strengthening of peasants' rights and collective action in the process. In facilitating land transferability and scaled agricultural production, it is undergoing a risky transformation that may backfire. My study reveals that this reform has caused social fragmentation, weak collective power of the poor and unsustainable natural resource use and farming practices. The Chinese approach to land tenure reform bears resemblances to other countries whose experiences have largely failed the poor and have produced unintended consequences. In essence, the failure to take into account the livelihoods of the poor, especially from sustainable land use perspectives, exemplifies the pursuit of short-term gains rather than longer-term solutions to complex rural development issues.

Although land tenure is important to sustainable development, it is not the only contributing factor. A particular land tenure system can only work in the long run provided that the overall biophysical, social, political and economic conditions support it. Thus, land users ought to be given the choice and discretionary power to define their preferred land tenure systems with the strong support of government, businesses and the wider public. Unless the land tenure system addresses the wider determinants of institutions, power, politics and social development, poor peasants will continue to remain marginalised in their struggles to articulate their interests. It is important to provide more institutionalised space for the poor to participate in land governance. Thus, a holistic understanding of what kind of land tenure systems exist in China, how they have worked in the past, what their problems are, and how they can be redressed to suit the needs of local communities, is developed. The book delves into contributing to an enhanced understanding of the nature, dimensions and context of China's land tenure reform by emphasising the importance of linking policy into the conditions and dynamics of land tenure systems. It contends that no matter how the degree of land tenure security can be improved, a sole focus on the strengthening of land rights is insufficient in addressing the wider development and governance complexities that shape the viability of land tenure.

In using an interdisciplinary approach to the study of China's land tenure reform – past, present and prospects – it provides a critical lens to examine the conditions and dynamics of land tenure, rural development and governance linkages and the underlying social, political and economic context. It discusses the controversial history of China's land reforms to throw light on the political nature of the reforms. In a review of China's reorientation towards more individualistic and pro-market instruments in land policy and legislation changes, it outlines the institutional challenges for sustainable land use and management. On the basis of this framework, research was conducted on the ground where local land tenure practices and experimentations have taken place in both developed and poorer regions. It maps out different land tenure systems – individual, collective, shareholding and commune – as well as their impacts on the livelihoods of the poor, natural resources management and rural governance, local responses and local institutional innovation in land tenure arrangements. This approach takes land tenure as an integral part of rural development and village governance as interwoven with multiple social, political, economic and biophysical parameters.

This book is a major output of the Protection of Farmers' Land and Property Rights in China (ProLAND) project at the University of Gronignen, funded by the then Dutch Ministry of Economic Affairs during 2007–2010. This project provided an indispensable contribution to an understanding of China's land tenure reform and land management through research, training and international exchanges for Chinese land researchers, experts and policy-makers. Our collaboration with Chinese and Dutch partners was stimulating and fruitful. As the then-manager of this project, I owe much appreciation to the wide support of the entire project team and our funder. I would like to extend my sincere gratitude to those land researchers, specialists, and especially the local government staff and villagers in my research sites for their ideas, insights, advice and kind facilitation of the research process.

At the University of Groningen, this book has received the kind support of my colleagues. In particular, I extend special thanks to Peter Ho and Leon Verstappen for their incessant efforts in supporting my research and work. I benefited from their insightful comments and advice. I should also thank Pieter Boele van Hensbroek, Leandro Vergara Camus and Hans Schoenmakers for their intellectual support and helpful advice especially at the early stage of research design and implementation. My appreciation also extends to the colleagues working at other Dutch universities: Benjamin van Rooij and Meine Pieter van Dijk for their constructive comments on the earlier drafts. I also owe my gratitude to the external anonymous peer reviewers for their crucial comments and suggestions. My appreciation is extended to Thomas Sikor and Jesse Ribot for their endorsements of the book, and to Nienke Busscher for her kind administrative support

Since early 2011, I have been coordinating a new research project titled 'Farmland Acquisition and Governance in China: Participatory Learning and Experimentation (LANGPLE)'. This project delves into the critical issues of transparency and accountability in farmland management processes by seeking to understand the institutional constraints to stakeholder participation. Challenging as it is, it is a timely contribution into the land policy reform with an increased emphasis on safeguarding the rights of those who have to give their land up for development, by providing the needed institutional analysis and methods of stakeholder participation. The design of this project was made possible by the findings and thesis of the book; dialectically, the completion of the book has benefited from the research part of the project. I thank my Chinese partners Weichang Li, Xiaohu Huang, Chujun Feng, Yibin Lu, as well as the programme officer Kate Hartford, and many other Chinese and international land experts, for all their strong interests and support for the project .

Working with the publishers Ashgate has been an enjoyable process. My special thanks are extended to Katy Low, Val Rose, Kayleigh Huelin, Pam Bertram, Carolyn Court and the entire editorial and production team for their patience, hard work and helpful hands-on support to ensure quality production of the book. It was a great honour to work with Ashgate as a world leader in publishing quality academic works.

Last but not least, I hope that the reader will gain some insights from this study. I must say that it remains an uncompleted mission, as I need to spend more time in the field to better understand the nuanced cases of land tenure, development and governance linkages. In this endeavour it remains a huge challenge to obtain the needed findings as access to the field is firstly exceedingly difficult for any researchers in the Chinese local context. The challenge of doing more in-depth fieldwork sets a major constraint to our understanding of the complex issues. I hope that as China's land tenure reform develops, supposedly towards greater land management transparency, more space can be opened for field research and multi-stakeholder engagements to serve the needs for the improvement of pro-poor development policies and institutions.

Yongjun Zhao
Assistant Professor of Globalization Studies and Humanitarian Action
University of Groningen, the Netherlands

Chapter 1

Introduction:
The Mystic Role of Land Tenure Reform in Sustainable Rural Development

Land Tenure Reform in Contextual Perspectives

While the role of land tenure in securing citizens' rights and access to landed resources is widely recognized as essential to sustainable livelihoods and development in any country context, ongoing reforms of land tenure in many countries have not proven it is the case, given the failures in the enforcement of land rights – be it individual, group or communal. At least, one may argue that this is simply because of the insufficiency, inefficiency or inherently problematic design of the reforms. If this holds true for those countries that are struggling to get their institutions right, China is no exception, but is unique in offering the rest the crucial lessons learnt in its own success and failure during the course of economic and social transformation. With a population of 1.3 billion, China still has more than 23 million people who do not have adequate access to food supply and shelter, and more than 40 million people living on an annual income of less than US$140. As the second largest world economy and at the same time a lower-middle income country with a rising income disparity between the rich and poor (World Bank, 2007), one may wonder what the role of the land has been in an economic growth underpinned by the daunting challenges of poverty and inequality in both rural and urban areas.

Emanating from its market reform in the 1980s, China's remarkable economic growth has been coupled with substantial loss of its natural resources, upon which the majority rural population rely. In recent years, the loss of arable land has reached almost 700,000 ha annually. Between 1987 and 2001, conversion to non-agricultural land use rendered at least 34 million Chinese peasants landless. It is estimated that by 2030 the total number of landless peasants will exceed 78 million (Zuo et al., 2004: 116–17). This vulnerable group, especially in poorer regions, has found it extremely difficult to pursue other economic opportunities to make ends meet. While rural–urban migration might offer a solution as promoted by government policy, poor public service provision in Chinese cities and rising urban population and poverty have made many migrants return to their villages of origin, especially with the advent of the financial crisis in 2008. Other factors such as climate change, drought, soil erosion, desertification and downgrading of farmland fertility have further constrained the government's goal of sustainable development and social harmony. The side effects of economic growth have serious repercussions on agricultural sustainability in general and national food security in particular. Keeping the remaining 1.8 billion ha of arable land intact has been at the top of central government policy, as seen for instance in the 12th Five-Year Plan (2011–2015). Otherwise, China has to rely more on food imports and development of a grain base in other continents such as Africa, for China has been a net food importer since 2003.

Will the ongoing land tenure reform focusing on the strengthening of landownership and use rights under market principles and mechanisms, preconceived as essential to create the right incentives for peasant households, help solve the above problems? This overarching question is the central theme of this book. Before putting land tenure reform into historical and contemporary perspectives, which will be explored in subsequent chapters, it is important to throw light on its multifaceted dimensions.

In the vast Chinese countryside, with an average of 0.08 ha of arable land per capita, not all the peasants have a shared strong interest in land investment due to high capital costs and extremely low economic returns from farming.[1] Many of them are not willing to receive their land use contracts and certificates due to their concerns over the heavy land-related taxes and fees imposed on them.[2] It is not even rare to find that frequent land adjustments are made on a regular basis to accommodate demographic changes, especially in poorer villages. Any reduction in the number of family members in a household would lead to the loss of land parcels to those families with increased members to balance out their needs for land, albeit this practice often causes conflict among the peasants and is forbidden by law. Women are vulnerable to the loss of land upon marriage and divorce. Although the law grants peasants renewable 30-year use rights, the peasants do not know how to use the law to resist sporadic evictions by government for developers (Zhao, 2008).

Furthermore, as farmland primarily serves the purpose of eking out a living with growing economic, social and biophysical constraints for most peasants, out-migration provides a temporary alternative. As a result, women, children and the elderly are usually left on the land struggling with very mean support from their relatives in the cities. In this context, many peasant households are not willing to perform land sub-leasing among themselves given the extremely low economic return from the land. Some abandon their land rather than transfer it to others when they migrate to cities under the assumption that they can still use it upon return to the village (Zhao, 2008).

---

1   Peasants' low interest in farming can be conducive to land loss especially in peri-urban areas where the value of land is much higher than that in rural areas.
2   Although the abolition of agricultural tax policy began nationwide in 2006, Chinese peasants are still the bearers of heavy economic burdens with increasing costs of living, materials, education, health and so on, all of which have repercussions on their choices over land use.

Village collectives, the lowest level of the quasi-state representative, together with local government and business enterprises, pursue common interests in land-related rent-seeking. According to a national survey, more than 50 per cent of land acquisition projects are intended for road construction, 16 per cent for factory, and 13 per cent for development zones or industrial parks (Zhu et al., 2006: 781). Local governments have greatly relied on these proceeds, which constitute two-thirds of their extra-budgetary revenue on average at the national level to meet economic growth targets, since they receive inadequate fiscal transfers from central government. But land evictees do not receive proper notice and they are unable to voice their concerns over unjust compensation and resettlement arrangements through the courts and other official channels. Their vulnerability in these struggles to gain equitable access to land use is also exacerbated by a lack of effective organization due to the widening economic and social division among themselves. The most eminent form of village organization – the regular village congress, for instance – is often bypassed because the village collective has less capacity to rally the masses than it did in the past due to inherent economic and political problems. In a study of selected villages in Sichuan, one of the poorest provinces, devastated by the massive earthquake in 2008, it was found that around 30 per cent of peasants do not participate in village elections, and 70 per cent of them do not know how to deal with corrupt leaders (China Institute for Reform and Development [CIRD], 2001).

Obviously, rural governance is the key to equitable and sustainable development, and land use in particular. In 2006, for the first time the State Council Leading Group Office for Poverty Alleviation and Development launched community-centred rural development programmes with a view to piloting a novel participatory model of rural development in 60 villages nationwide that would involve civil society organizations (CSOs) in implementation. Local ownership was highlighted as the key to the success of these programmes. Deemed as a prominent shift away from top-down approaches to development, this model was intended to return power to the people who could have a stronger voice in project design, implementation, monitoring and evaluation while integrating poverty alleviation into village self-governance and democratic decision-making.[3] The wide range of fields covered included community development, health, education, water resources management and agriculture, and so forth.

In fact, the exploration of more effective rural development measures has always been on the political agenda of the Chinese government. The post-1978 economic reform agenda, aimed at replacing collective agriculture with

---

3   The State Council Leading Group Office for Poverty Alleviation and Development is the principal department in charge of policy-making and implementation concerning poverty alleviation. In recent years, it has collaborated with the major international development organizations in piloting community-based innovative development projects. This latest initiative also received the support of international donors such as the Asian Development Bank (ADB) and UK Department for International Development (DFID).

the Household Responsibility System (HRS), has offered peasant households substantial economic autonomy as the basic unit of rural agricultural production. Seen as a crucial step towards the revitalization of the rural economy in the aftermath of the collectivization era, the HRS was expected to bring about the needed changes in village governance through village elections for better rural development outcomes (Plummer and Taylor, 2004).

To a certain extent, the contribution of the HRS to equitable and inclusive development cannot be overestimated, as seen in its limited ability to ensure that poor and disadvantaged households can claim their rights and voice their concerns over village affairs. In this sense, the participatory model under experimentation can perhaps be understood as a supplement to the HRS and a last resort for government to win the support of the peasantry. But how it can have the expected effects when the HRS and village elections themselves have not been effective in poverty alleviation and participatory village governance? In other words, how can this model address the issue of power and agency in the wider rural development landscape? And how can greater poverty alleviation outcomes be achieved when there is lack of genuine grassroots governance? (see Xu, 2003; Hutton, 2006; Pei, 2006).

These questions concern the generic question of what village governance mechanisms can contribute to more effective poverty alleviation results, which cannot be answered easily without substantial empirical research. But at least village governance is interwoven with the wider political economy of the village and country as a whole. A way to fill in this research gap is to locate land as the most obvious factor and means of production embedded with peasant livelihoods and village governance. Rural development programmes that have not paid due attention to the role of land in rural political economy may even backfire (see Sargeson, 2004).[4]

The land reform history of the post-1949 era reveals that land has been a crucial impetus for the overall economic and political reform agenda of the government. According to Xu (2003), once land and other natural resources were institutionalized as publicly-owned, the government automatically managed to put society under its direct control. Yet this does not mean that there is no margin for peasant-centred rural governance. The substitution of the HRS for the commune system has enabled the creation of relatively autonomous family groups, whose rights are sometimes in conflict with those of the public or nation-state. This struggle for household interests should be seen as a fundamental basis of grass-roots democratic governance.

Given its strong correlation with agricultural inefficiency and chronic rural poverty, the HRS has been widely attacked by the neo-liberal school, however. For them, it is a system of household farming under the direct management of the

---

4   Land is a sensitive issue for many government departments and international development organizations, which, except for the World Bank, UNDP, DFID and Ford Foundation, have hardly embarked on land projects till now.

village collective or administrative committee, as rural land is collectively owned by law. This dual institutional arrangement has been seen as a stumbling block to market-oriented agricultural development and rural governance, for the dominant force of the village collective can jeopardize the participation of the majority poor in land and agricultural management (see Wang, 1999; Xie, 2001; Chi, 2002). Nonetheless, it is unlikely that the government will change its current course in the foreseeable future. This further indicates the government's conservative approach in order to maintain social and political stability. The inextricable link between rural development, governance and land reform remains to be further understood for innovative economic, social and political reform.

In a nutshell, given an extremely low availability of arable land for basic livelihoods, how to ensure that the peasants can maximize their benefits from the land remains a critical challenge (Wen, 2005). Continuing landlessness will exacerbate the increasing social inequality and impoverishment of poor and disadvantaged groups. As the land struggles loom large, the Chinese peasantry may be put at the mercy of the state which has yet to come to grips with a sound strategy of institutional change.

The government's efforts in curbing illegal land evictions and disorderly land management practices are evident in many policy circulars and revised laws which have one thing in common – imposing more stringent rules on local government behaviour to improve accountability mechanisms with, however, few incentives for peasant participation. A key policy dilemma closely matches the huge difficulty in balancing stakeholder interests and socialist and market mechanisms for land tenure arrangements. Concerns over social and political instability override any attempt to institutionalize land privatization. Ironically, establishment of a fully-fledged land market is increasingly deemed by many experts as the last resort to facilitate free and fair land transactions without overt government arbitration. Nevertheless, the land sector is exemplary of the socialist market economy in which the state needs to put everything under its strict control and surveillance. Concerned about the trajectory of land policy changes, a renowned scholar from the State Council Research Office commented:

> We all know the problems, but linking them with policy is another thing, because when you change one policy, then it is hard to predict its unintended consequences. That is why the current land policy is focused on stringent administrative measures instead of those for land privatization.[5]

Addressing the central theme of the book requires a deeper understanding of the social, economic, political and biophysical dimensions of land tenure in order not to oversimplify the latent constraints to land tenure arrangements. Although the concerns about land tenure insecurity and weak property rights are reasonable, this does not render easy solutions such as the institution of land privatization. The

---

5   Informal interview in April 2009.

latter, for some, is presumed to enable peasants to sell and buy land and eventually develop it into large-scale farms, which can benefit both agricultural productivity and rural development, and provide a firmer establishment of the rule of law and democracy in the Chinese countryside (Mao, 2003). However, as Wen (2005) contends, what then would happen to those hundreds of millions of displaced subsistence peasants? International experiences also show that there is no definitive relationship between land tenure and peasant investment (Dekker, 2001). Peasants are more preoccupied by political and economic insecurity than insecure tenure or land title. Policy-makers should focus more on the rural sector and broader judicial and political reforms rather than tinker with the tenure system. To a certain extent, the current land tenure system in China has ensured a social safety net for the poor and avoided the growth of a large landless class as seen in many other developing countries (Huang and Pieke, 2003). It is more urgent to address poverty, inequality and agricultural development than to drastically privatize the land, and it is more important to improve legal access for the rural populace to enhance their social and economic rights as a precondition of any land tenure reforms.

'Reinventing the Wheel': The Land Tenure Question

Obviously, land tenure and its relationship with village governance and sustainable rural development deserve an in-depth study. The current land tenure system in China has not been analysed holistically, particularly in terms of its social and political dimensions and the linkages with sustainable natural resource use and agricultural development. Some optimistic views on the contribution of land tenure to economic and social development cannot hold true for the inherent need for wider institutional reform. As Rigg (2006: 198) argues, there is a need to reconsider some old questions (e.g. teleological thinking on landownership and economic development) on how best to achieve pro-poor development in the rural South, as livelihoods have become de-linked from farming, and poverty and inequality arise from landownership (see also Bandeira and Sumpsi, 2009). China bears resemblances with this case, for urbanization continues to undermine the possibility of sustained rural development and agriculture in particular. There is a need to draw on more empirical research into how land tenure is perceived by different stakeholders and how it is organized by the people themselves in order to develop better understanding of the multi-faceted nature of land tenure and the conditions in which appropriate land tenure systems can be explored.

My study aims to testify that land tenure regimes, be they collective, customary or private ownership, are inextricably linked with the economic, natural resources, social, cultural and political conditions of a given setting. Major challenges for policy-makers are to delineate and work out the parameters for changes in inter-related historical, social, political and economic processes among different actors at different social organizational levels. A proper fit among these dimensions determines the security, appropriateness and effectiveness of land tenure. Thus, a new paradigm shift needs to be developed towards a more locally-fit approach that is participatory in nature and grants the peasants the choice over their preferred land tenure systems. A land tenure regime can only work in the interests of the poor and disadvantaged groups when it is compatible with the other conditions – the aggregate effects serve the sustainable development need of the poor population. By doing so, this book will also contribute to the forecasting of China's agrarian future underpinned by the rising contradictions of urban and rural development and sustainable natural resource use and governance.

The ultimate goal of the book is to contribute to the understanding of land tenure reform and local practices in China and the underlying institutional changes needed to tackle land-related poverty, power and politics. Embedded within this framework is an exploration of locally-experimented forms of land tenure systems as potential alternatives to the conventional models. The study consists of three specific objectives:

- to contribute to the theoretical development of land tenure and property rights approaches;
- to develop a critical understanding of land laws, policies and institutions underpinning sustainable land use and China's social, economic and political transformation;
- to provide insights into the opportunities and constraints for land institutional experimentation at the village level.

These three thematic issues lead to the identification of possible solutions to current policy inertia and failures in farmland preservation and sustainable land use. Although there are no immediate answers or one-size-fits-all solutions, my study reveals the need for an exploration of flexible approaches to addressing the pressing issues of tenure insecurity, poor governance and poverty in the Chinese countryside. The following specific questions are discussed to this end:

- What historical, social, political and economic contexts can inform land tenure reform in China?
- To what extent has the current land tenure system facilitated or obstructed village governance and sustainable rural development?
- How are different land tenure arrangements linked with land use, rural governance, social and political relations?
- What are the social, economic and political meanings derived from the land reform process for different actors, how are these meanings construed and what are the implications for formally and informally organized land use and management practices and power relations among them?

Research Approach and Constraints

To provide a snapshot of the key governance and social development issues around China's land reform as an ongoing process with regard to land tenure, rural governance, land use and development linkages, illustrative cases derived from different parts of the country are used to serve the contextual bases of policy, legal and institutional processes at central, regional and local levels. These cases are focused on farmland acquisition, use and management, all of which are underpinned by poverty and inequality as the key priorities for government's redressing. In particular, it draws on two fieldwork cases in contrasting regions – Hebei and Guangdong provinces in terms of their disparate economic, environmental and social parameters. Hebei is chosen as one of the poor regions that are experiencing depletion of natural resources, chronic poverty and unsustainable land use and governance. In particular, it exemplifies the issues around the implementation of the HRS in agricultural development. By contrast, Guangdong is one of the most developed economies and the first one to enjoy the open-door policy started in the early 1980s. Local governments here have taken bolder action than in Hebei in bringing forward institutional experimentation in varying forms with, however, many problems being encountered. More importantly, in many peri-urban areas of this province, many peasants have gradually lost interest in keeping the land for farming, as they would rather give it out to local developers in return for high cash compensation.

Ethnographic methods were applied in combination with other qualitative approaches given the interdisciplinary complexity of the research topics, and the need to look into the perspectives of different social and political actors. Primarily, participant observation was used to allow a great level of flexibility in adjusting different methods and engaging with a wide range of actors in the research process (see Lévi-Strauss, 1976). By doing so, the questions concerning the ways of life of the community, their social interactions and perspectives were explored. Moreover, in working with different stakeholders – farmer households, business enterprises and government officials, this study emphasizes the importance of interactive and dialectic processes of engagement. This method is quite useful in the context in that probably the interviewees know much more about the researched issues than the interviewer, and building trust and equal partnerships is the key. In essence, the approach was to gain access to information and the views of the stakeholders through informal means rather than formal interviews. It was a learning-by-doing process in which the research questions simply served as a broad scope of guidelines. The stakeholder views were cross-checked at different locations and jurisdictions in terms of similarities and differences, which were translated into data for systematic analysis. Thus, not only was this approach not confined to the study of community at the micro level, but was also useful for macro-level analysis of powerful groups and institutions and their interactions.

The selection of specific research sites became a lengthy process due to the necessity of reaching out to many contacts in China. Without their support it would not have been possible, especially for a Chinese researcher working at a Dutch university. The collaboration with research institutions in China enabled direct contacts with some government staff and village cadres. This greatly facilitated field entry. By doing so, it was possible to 'study up' in terms of arranging meetings with government staff and obtaining their views on related topics. Despite the efforts made in approaching the interlocutors, difficulties were met in making appointments and having conversations with some contacts. Their reluctance to receive me also posed a challenge to the whole research. There was little chance to hear their critical views on relevant policies and laws. It was reasonable that they had many concerns, but they were also quite selective in the information they were willing to provide. As the research topic is complex in nature, it was hard for anyone to provide informative facts, figures and viewpoints that could be of direct use.

The fieldwork was the most difficult part of the study. Even with local government and village cadres' consent, it was a huge challenge to touch upon those difficult research questions. A lack of support from the peasants during the fieldwork was a major constraint, for their varied concerns over the unexpected outcomes of their answers deterred their active participation in the research process. It was not possible to just enter the homes of randomly selected households for an interview even with an approval letter from the local leaders. Some households showed strong reluctance to cooperate. This constraint nullified the use of the conventional methods of questionnaires and focus group discussions.

To overcome this constraint, building trust was crucial to the entire fieldwork process, but not enough. I experienced this difficulty, for instance, in the paddy fields talking to the tillers. In many cases, although they could not drive me off the field, our conversations did not last as long as I had hoped. But the more people I talked to, the easier it was to at least find those who were interested in the talking. This is not only a matter of trust, obviously, but shows how diverse the groups with disparaging perspectives on the research questions were. A major lesson learnt is that the researcher should think and act ahead with an intention to share the feelings of the local groups and have a passion to extend a helping hand especially to the elderly, women and children. The more time spent in the field just observing what was happening, the better the local conditions were understood. This supported further interactions with the people on their livelihood practice, life histories, activities, organizations, networks and viewpoints (see Mitchell, 1969). All the findings obtained from the field were thus combined with those from the government and research institutions for further analysis.

In a nutshell, I took a flexible approach to the fieldwork and meetings with government and research staff to develop a reasonable degree of trust, openness and honesty, which was crucial for the validity of the data and continued research with these participants. During the process of research analysis, for instance, I managed to keep the key informants updated of my research progress and continued to solicit their views on the findings. Moreover, they continued to provide me with updated information in this field. This way of research was a mutual learning process, in which I benefited from the experiences and views of the informants who contributed to the finalization of research data analysis.

Desk research into the existing literature on land reform in China was conducted throughout the project, which was used to reflect upon my research findings. However, there is limited literature on China relevant to the research topic, although it is useful in certain ways. In particular, empirical studies on micro-level land problems remain minimal, which is a key constraint to the development of a critical mass in debates on local views and practice. This disadvantage was dealt with by actively resorting to the literature on international experiences and practices in relevant subjects, and linking them with the case of China. This effort played an important role in defining the research topics highly pertinent to China, which needs to draw serious lessons from other countries.

Overall, the empirical findings collected serves the purpose of the analysis of different stakeholder perspectives rather than quantitative analysis. However, due to the above-mentioned constraints to the fieldwork, it was a huge challenge to gain much needed in-depth study of the local situations, which remains a goal for the furtherance of this study.

Book Structure

To give the reader an integrated and comprehensive overview and critical analysis of the issues, debates and findings, this book comprises stand-alone but interconnected article-type chapters. Each chapter provides a context for the others and enables the reader to find their linkages from a more contextualized and critical point of view.

Chapter 2 is the theoretical construction illustrative of the major schools of thought surrounding land tenure reform and its underlying challenges of sustainable development and governance, which call into question the preconceived wisdom on land and property rights. Chapter 3 contains an historical account of land tenure changes especially since the late Ming dynasty, characterized by the emergence of capitalist productive relations to provide a wider spectrum of the nature of China's land reform and its effects on China's rural society. It demonstrates the importance of power and politics in the reform process and its missing link with the rural reality – demands of the poor for the forms of land tenure that suit their best interests. As a follow-up, Chapter 4 provides an up-to-date overview of major land policy changes since the market reforms took stage and their trajectories and impact on the livelihoods and social and political relations of the rural poor and disadvantaged groups. It suggests more inclusive approaches to China's land policy reform rather than the simplistic market-oriented approach that overestimates the role of the land market for good governance and sustainable rural development.

Against this backdrop, three studies of local practices in land tenure arrangements are followed to further illustrate and explain the major contentions developed in this study. Chapter 5, based on the fieldwork in Hebei province, northern China, presents a case of the current land tenure regime – the HRS in agricultural development and natural resource management in a poverty-stricken region facing severe environmental challenges. Focusing on the concomitant effects on the lack of diversified livelihood strategies of the poor, and land degradation, it delves into the fragmentation of social and political relations in the rural community as a severe challenge to community-centred participatory approaches to land tenure. It shows that a land tenure system cannot be sustainable if rural development and good village governance are not coupled and supportive of its existence. Furthermore, as Chapter 6 demonstrates, the so-called innovative local practices in land shareholding cooperatives do not actually constitute the collective institutions of the weak groups who lack voices and power to hold these institutions into account. By exploring their effects on the peasant land shareholders and intrinsic problems of poor governance and inappropriate land use and management practices, it argues that the land tenure system should be treated as a dynamic process where local stakeholders continue to formulate it and put it to the test. Chapter 7 is based on fieldwork in Guangdong province, southern China, and shows the unique case of a commune village and explains why and how it has managed to survive the infiltration of the mainstream market political economy in general and land takings in particular. It further demonstrates the role of collective power in articulating their preferred tenure system and managing it in their interests. It is an institutional demonstration at the village level of what constitutes a feasible land tenure system for the poor whose notions of land tenure may differ drastically from those of many land experts and decision-makers.

In conclusion, Chapter 8 reviews the major theoretical issues on land tenure, property rights, institutional change and rural development informed by this study. It summarizes a critical framework for the analysis of land tenure reform in China, which serves as a stepping stone towards the development of relevant theories on a wider scale. Moreover, it crystallizes key policy issues surrounding land governance improvements and institutional arrangements needed to tackle the fundamental challenges of rural poverty, social inequality and unsustainable landed natural resource use, all of which have repercussions on the design of sustainable land tenure systems for the poor and disadvantaged groups during China's transition.

References

Bandeira, Pablo and Sumpsi, José María. 2009. "Access to land, rural development and public action: The when and the how", *Development Policy Review*, 27 (1), 33–49.

Chi, Fulin. 2002. "WTO accession will accelerate reforms in China", CIRD, www.chinareform.org/cgi-bin/researchpaper

China Institute for Reform and Development (CIRD). 2001. "重建农民对土地的信心 – 四川农地制度现状报告" (Rebuilding villagers' confidence in land – report on the status quo of rural land system in Sichuan), *CIRD Bulletin*, Vol. 364, 22 November.

Dekker, Henri. 2001. *A New Property Regime in Kyrgyzstan: An Investigation into the Links between Land Reform, Food Security and Economic Development*, PhD thesis, University of Amsterdam.

Huang, Ping and Pieke, N. Frank. 2003. "China migration country study", paper presented at the conference jointly organized by the Refugee and Migratory Research Unit and the Department for International Development, 22–24 June 2003 in Dhaka, Bangladesh.

Hutton, Will. 2006. *The Writing on the Wall: Why We Must Embrace China as a Partner or Face It as an Enemy*. New York: Free Press.

Lévi-Strauss, Claude. 1976. *Structural Anthropology*, Vol. 2, New York: Basic Books.

Mao, Yushi. 2003. "改革的动力在党内" (The driving force for further reform resonates with the Party), *Gaige Neican*, 2003 (11), 19–21.

Mitchell, J.C. (ed.). 1969. *Social Networks in Urban Situations*, Manchester: Manchester University Press.

Pei, Minxin. 2006. *China's Trapped Transition: The Limits of Developmental Autocracy*, Cambridge: Harvard University Press.

Plummer, Janelle and Taylor, G. John. 2004. *Community Participation in China: Issues and Processes for Capacity Building*, London: Earthscan.

Rigg, Jonathan. 2006. "Land, farming, livelihoods, and poverty: Rethinking the links in the Rural South", *World Development*, 34 (1), 180–202.

Sargeson, Sally. 2004. "Full circle? Rural land reforms in globalizing China", *Critical Asian Studies*, 36 (4), 637–56.

Wang, Xiyu. 1999. "在家庭经营基础上深化农地制度改革" (Deepening the reform of the rural land system on the basis of household management), *China's Rural Economy*, 1, 4–8.

Wen, Tiejun. 2005. 三农问题与世纪反思 (*Reflections on the three rural issues at the turn of the century*), Beijing: Sanlian Shudian.

World Bank. 2007. *World Development Report 2008: Agriculture for Development*, Washington, DC: The World Bank.

Xie, Junqi. 2001. "土地" (Land), 中国环境与发展评论 (*Review of China's Environment and Development*), Beijing: Social Sciences Academic Press (China).

Xu, Yong. 2003. 乡村治理与中国政治 (*Rural Governance and Chinese Politics*), Beijing: China Social Science Press.

Zhao, Yongjun. 2008. "Land rights in China: Promised land", *The Broker*, No. 7, 7–9.

Zhu, Keliang, Prosterman, Roy, Ye, Jianping et al. 2006. "The rural land question in China: Analysis and recommendations based on a seventeen-province survey", *Journal of International Law and Politics*, 38 (4), 762–839.
Zuo, Ting, Li, Xiaoyun, and Ye, Jingzhong. 2004. "2003–2004 中国农村资源与环境状况" (2003–2004 Status of rural resources and environment in China), in Xiaoyun Li, Ting Zuo and Jingzhong Ye. (eds), *2003–2004 Status of Rural China*, Beijing: Social Sciences Academic Press (China).

Chapter 2
International Context of Land Tenure Reform: Theoretical Reconstruction

Introduction

As an emerging world power and the world's second largest economy, China has been a hotspot of both applause and criticisms with regard to its way of doing development both at home and more abroad. On the latter, China is often cited as a crucial contributor to all kinds of things one can imagine of, most of which relate to its adverse impact on local economy, livelihoods and natural resource governance, for instance, in Africa. Conversely, China's economic achievements have also been acclaimed by its African partners for its unique approach to development, which some African states may find it worthwhile comparing and if possible imitating, as the Chinese development aid to Africa has beaten those traditional donors in terms of the aid volume and diverse approaches. Nonetheless, it remains a puzzle as to whether there is indeed a China model that other countries can follow, as China is also facing severe challenges of social inequality, poverty, natural resource shortages and depletion, all of which set constraints to its agenda on building a harmonious equitable and sustainable economy. Social conflicts have been on the rise as a result of ineffective governance measures to safeguard the interests of the disadvantaged groups when local governments pursue the target of high economic growth rate at the expense of societal rights and interests. In the new 12th Five-Year Plan (2011–2015) for national development, for the first time social management is mentioned to coping with rising social unrests and discontents over the disparaging effects of economic growth.

At least, it can be seen that the China model is controversial. But what is certain rests with the stronger role of the state in pursuing its economic and political agenda as compared with that of African states given their differences in political economy. Paradoxically, as decentralization continues, the role of the national government has dramatically weakened, and sub-national governments have gained more discretionary power over local economic development. To a certain extent, central policies and laws are not implemented properly by their local constituencies. As a Chinese 'saying' goes, central political orders cannot get out of *Zhongnanhai* (often referred to the State Council). There are many reasons for this relationship, one of which is the inflexibility and inappropriateness of policy imposed on the local government. Chinese ministries produce many policy circulars and regulations trying to exert control and maintain order, which local governments do not take in a serious manner. These internal policy constraints concerning China's goal to achieve greater development thus contradict the grand views of those who advocate the China model for other developing countries. As land cross-cuts other development sectors, understanding its changing role in economic transformation in China can shed light on those debates pertaining to the trajectories of economic and political reform in general, and land reform in particular – not only in China, but also elsewhere. Although differing in country context, commonality can be drawn between China and other developing countries as both pursue the market-oriented 'magic' to fix the improper institutional arrangements deemed as obstacles to rapid economic growth. How to move land reform forward to facilitate this market transformation process as the economic reform deepens thus remains crucial in their policies. It is in this background that this chapter highlights some of the major schools of thought on land tenure reform with an analysis of their strengths and weaknesses in an attempt to challenge the conventional wisdom and propose alternative thinking on its problematic.

The Power of the Western Economic Model

While each country should develop its unique path to sustainable development and social and political institutions, entry into the WTO has meant that there is little leeway for China's own institutional innovation, especially in economic reform. Its fast economic growth, to a certain degree, has much to do with the unleashing of the land market that has penetrated the shrinking farmland. How to maintain the current level of farmland for the sake of national food security and social stability undoubtedly constitutes a top priority. Despite all the efforts made, how to make policies, laws and institutions more effective in regulating the market and thus enabling free and fair market transactions, for instance concerning the real estate bubbles over recent years, remains largely a puzzle despite recent stringent measures to curb inflation, which have made only minor headway. While the latter issue is not purely economic, the government seems to rely more on market-led approaches to addressing the draconian non-economic problematic. In this sense, at least, sound social and political measures underpinning economic reform have not been innovated; as a result, China's institutional reform is in stalemate.

There is a lack of analysis of the vacuum of institutional innovation at grand policy level, although local governments take the flexibility to test relatively desired approaches to suit their own interests, which might have effects on policy changes at the national level. Overall, this gap might explain the lack of clear direction (other than the call on the advanced stage of socialist market economy as the ultimate goal) in the long-term process of China's transformation. If the latter is not well defined or interpreted, it would be easier just to follow the market mechanisms in the short run.

> There is no unique Chinese model as such; in the end the ideal situation for
> China would be what you can see in the West with regard to a fully-fledged and

transparent land market and the economy as a whole. Of course, we cannot say this abruptly at the moment, but in fact, we are doing that.[1]

The fast economic growth China has achieved so far is often perceived by the West as a result of its state-predominant model of development (Hann and Hart, 2011: 103). Since the founding of the People's Republic, the Chinese government has been caught in a left or right syndrome with regard to its direction of the reform. But in general, it has favoured the role of the state in steering economic development, which shares many similarities with the modernization approach. Gradually, since the early 1990s, the market has gained momentum in economic policies, which can be taken as part of the influence of neo-liberalism and the failure of the planned economy. However, a sole focus on the magic power of the market has proven not fully desirable, as reflected in the current policy agenda on sustainable development and social harmony in reaction to rising social inequality and environmental degradation as the market economy entrenches (see Hann and Hart, 2011). The current financial crisis has hardened the government's choice over the needed institutional change and rising tensions within its party ranks. In the land sector, policies do not display any explicit inclination to furtherance of clarification of the collective landownership or land privatization, despite scholarly attacks on its ambiguity and links with tenure insecurity and social conflicts (see Ho, 2005). Apparently, only an incremental approach can be thought appropriate.

Pragmatically, some land officials and experts have shown keen interest in how the West has developed its property right systems concerning land registration and cadastral management over time and space, which may have useful lessons for their own policy improvements, for instance. At the same time, they have also been wary of the problems of the Western approach to land reform as experienced in other countries, especially transition economies. At least, they try to avoid the unintended consequences of land reform practiced in other countries. But perhaps they have paid more attention to those technical aspects of land administration rather than making headway in institutional arrangements for inclusive, transparent and accountable systems.[2]

It is known that the Western property rights approach provides legal binding status for individual property rights. Its core lies in the stipulation of property rights as an individual right enshrined in civil codes and constitutions. Moreover, land transfers are conducted based on a valid legal ground or a real agreement essential to any transfer, and land registration is institutionalized to reflect the high level of equity and transparency in land administration. Although countries differ in their legal stipulations in the transfer of properties, for example, in the cases of German,

---

1   Informal interview with a Chinese land scholar in Beijing in 2009.

2   This is evident in increasing research collaboration between Chinese and Western scholars in land administration and governance as well as Chinese policy-makers' engagements with the West to learn from their history and current practices of land policy and administration.

French, English and Dutch law (van Vliet, 2000), all these systems require every transfer to be backed by real agreement. In Dutch law, ownership is defined as the most comprehensive real right to the owner. Transfer of landownership requires the drawing up of notarial deeds, which are registered with a special public registry or the Dutch Cadastre. On common ownership, the Dutch Civil Code stipulates that the ownership of each owner of the common property must be written into a notarial deed and the deed must be registered (Wang, 2006). With regard to usufruct right, the person with this right can exert his right against anyone who infringes on it (Kleijn et al., 2006). In addition, many European countries are in the process of converting their land registry systems to electronic database format to allow electronic transmission of land transfer documents and direct access to the database to effectuate land transfers and registrations.

These developments have evolved over hundreds of years with the strong backing of political and social systems, with the ultimate goal of supporting the market economy, which can hardly happen in the short-run in the current Chinese and many other development country contexts. On these conditions, an overt focus on the market as a silver bullet will omit the fundamental governance changes which need to be coupled with the market reform. In other words, appropriate institutional arrangements where the market is just one part have to be made appropriate for the market economy and democratic governance (Hutton, 2006). The issue remains as to what precisely Chinese policy-makers can learn from the Western experience. The power of the Western economic model is assumed to have worked largely successfully in the West, but it is obvious that its adaptation by other countries remains problematic. Although the central government may attempt to standardize its legal and regulatory regimes, many localities may be reluctant to follow, or they may even resist the central rulings (Mertha and Zeng, 2005). It is difficult to foresee what measures the government will take to ensure that its rules and regulations are in harmony with local needs. This will constitute critical rules of the game underpinned by the entrenchment of the market economy.

Failure of the Western Economic Model in the South

In land reform, especially concerning land tenure and property rights, simply following the Western model of land titling and privatization has not proven a success for those post-socialist countries undertaking drastic land reform programmes. For example, Russia's land reform since the early 1990s with a view to de-collectivization has shown a rather low level of private farming and unequal access to land for poor farmers. The reform has reproduced the former Soviet forms of *de facto* property rights regimes and agricultural production. The majority of poor farmers still hold on to the collective means of production and are inactive in participating in the reform process, as they are marginalized by the powerful rural elite and local polity. Although a formal land transfer system is in place, it is the unwritten rules and informal procedures that have reinforced social stratification (Allina-Pisano, 2004). In Moldova, rural de-collectivization has brought unintended consequences in terms of extensive land fragmentation and the emergence of a land lease market whereby individual farmers lease their lands to agricultural enterprises, which in turn leads to land consolidation (Cashin and McGrath, 2006). In Vietnam, ongoing land reform with the introduction of systematic land registration has met with resistance from local communities because the new land rights imposed on them conflict with their actual land relations, and this reform has not brought about greater tenure security for farmers, nor has it exerted any major effect on agricultural growth (Sikor, 2006).

These cases illustrate the complexities and scales underpinning the functioning of property systems. As Ye (2000) points out, the reforms in these countries have favoured non-agricultural speculative groups which dominate the land market to gain lucrative benefits; as a result, many poor farmers became their tenants. In essence, the reforms aimed at ensuring equity and efficiency in the functioning of land management systems have actually triggered further disparity of social and political relations and deprivation of the disadvantaged groups in terms of land loss and inability to access basic social services.

Furthermore, African experiences of land reform illuminate the fact that land reforms in line with the Western model are inextricably linked with colonial history, local politics and culture, all of which have strong bearing on implementation failures (Daley and Hobley, 2005). Programmes underscored by the individualization of landownership are seen as a threat to social security as a result of enlarged land holdings and landlessness. Contrary to the reform goals of redressing tenure insecurity, claimed to be associated with customary land tenure systems, the latter are in fact found to promote a sense of communal responsibility for land resource management. Thus, Africa may need flexible alternatives to the existing statutory systems being tested, and that common rights models in the name of communal titles should be further researched rather than simply be replaced. From a technical viewpoint, land conveyances through land deeds and title registration have proven to be costly to the majority poor who cannot afford to pay relevant fees (Törhönen, 2003). As a result, land titling programmes have not greatly reduced tenure insecurity, but promoted social inequality, weakened the position of women and exacerbated landlessness. Recognition of the role of customary systems in land management is called for to redress this issue. For instance, in Botswana, customary land tenure systems and statutory law co-exist, which provides an innovative and robust land management system in response to societal needs (Adams, 2003).

International organizations aiming to introduce land titling programmes in developing countries have taken a precautionary approach. More community-based approaches are promoted to accommodate low-income groups. Their past experiences show that the forms of landownership depend on the nature of the resource itself and existing social arrangements. And an effective land policy reform will only be made more feasible by an open and broadly based policy dialogue, carefully chosen and evaluated pilot projects and sharing of experiences across countries (Deininger et al., 2003: 17). Notably, land titling programmes under certain circumstances can contribute to tenure security and improved welfare. But tenure security or the farmers' secured rights to use land can be achieved through other means than individual land titling and registration (Palacio, 2006). It is important to learn from these cases how formal policies and informal or local practices are interwoven and shape each other, the implications of which would be useful for furthering land policy reform.

Given the social and political complexity in land tenure reform in the South and the fact that Western models of land administration may not contribute to the effective functioning of a rural economy, there is a need to find ways to strike a balance between market-oriented approaches and state intervention in property rights arrangements. The state must play a key role in guiding institutional changes in the reform process (Ho and Spoor, 2006). Moreover, community participation is a prerequisite for land reform programmes to build up legitimacy for land administration in studying the societal needs and understanding complex local realities in which poverty, power and politics are interwoven with societal choices for viable programmes (McEwen and Nolan, 2007).

However, lessons from the South suggest that agrarian reform does not follow the transition paradigm. This paradigm, marked by a linear change from a traditional economy or communal system to a modern market economy with the introduction of more market and socially and technologically advanced elements, has proven an illusion. Rural opposition to land privatization suggests that market-oriented production is a social contract, which has to be built over time before achieving possible success (Ellman, 2003). Or simply put, there may be other alternatives to market-led approaches to land and economic reform, which remains unaddressed. In any case, land reform programmes should be context-wise to cater for specific social, economic, cultural and political realities.

Critique of Development and Legal Centrism

The process of land reform cannot be understood in isolation from a country's overall development context. Development can be interpreted as a social, economic, political and cultural process (Grillo and Stirrat, 1997). As part of this process, land reform is inextricably linked with social relations at various social strata; an understanding of these complex relations can provide insights into its underlying social and cultural issues. From this angle, a critique of development policy and practice will provide insightful perspectives on its positive and negative impacts on a given society. By doing so, an inquiry into the nature of local power and hierarchy, the nature of household and rural collectives relations, organization of local property relations and community organizations can be made. And the meanings of diverse discourses of dominant actors can be defined (Gardner and Lewis, 1996: 89; Rutherford, 2004). In other words, development is not all about bureaucratic planning with taken-for-granted rhetoric; rather, it ought to accommodate people's real interests and practices (Hann and Hart, 2011: 107). A shift of this approach to the study of China's development model in general and land reform in particular is needed to fill in the current gap in theory and policy practice.

As the success of land reform, to a large extent, is contingent upon appropriate institutional arrangements, there is a need to delve into how these arrangements can be made and how they function. Institutions are the rules of the game in a society or, more formally, they are the humanly devised constraints that shape human interaction. They structure incentives in human exchange, whether political, social, or economic (North, 1990: 3). North's definition is useful to understand how institutions are socially and culturally constructed. One needs to study social institutions as a system of patterned expectations about the behaviours of individuals fulfilling their socially-recognized roles. And institutional development should be focused on their working dynamics, discourses and contextual relations (Cotterrell, 1992; Lewis, 1999). Institutions are negotiated, contested and filled with multi-vocal discourses that need to be uncovered in a field of contestation (Abram, 1998). In the case of China, the existence of rural collective control of land may provide such working rules that enable peasants to move in and out of their land in response to changing economic conditions in the larger economy. There is a need to further look into these rules and seek more appropriate institutional arrangements that promote viable and productive agriculture in China (Bromley, 2005).

Furthermore, property institutions embody a bundle of interactive public and private rights. These rights coexist with other often contradictory and regulatory orders at different layers of social organizations. These organizations carry various bodies of cultures, ideas and ideologies, normative and regulatory institutions, layers of professional and day-to-day practices and everyday social relationships and actors' interests, which are referred to as legal pluralism. The latter poses a challenge to legal centrism which often misleads public policy and ignores the social context in which resources and property right regimes are embedded (Spiertz and Wiber, 1996: 13; Biezeveld, 2002: 11). To study legal plurality, one needs to understand the 'living law' that manifests itself in the principles abstracted from the actual behaviour of the society studied. 'What state considers to be "their land" is often defined as the land of individuals, families, lineages or communities by local or non-state laws' (Benda-Beckmann, 2006: 67). By studying the 'living law', the relationships between law and extra-legal aspects of culture can be further revealed (Pospisil, 1974). There is a need to examine how the current land law and other institutional arrangements work in practice, how they exert impacts on the local community and how the community reacts to them.

Land institutions should also be responsive to changing social conditions, which demand legal development. As Nonet and Selznick (1978: 14) put it, law acts as a facilitator of response to social needs and aspirations. To a certain extent, the Chinese laws are ambiguous, fragmented and sometimes self-contradictory, and are not equipped to cope with the changing needs of the poor (Ho, 2003).

To mitigate social conflicts over land, the state would need to be more proactive in dealing with the current problems. A study of law can be of importance in providing evidence-based policy recommendations on how the state apparatus should improve land policy through developing more responsive measures to cope with current constraints and to build up a fully-fledged legal system. It contributes to bringing about a reintegration of legal, political and social theory and recasting jurisprudential issues in a social science perspective (Nonent and Selznick, 1978). Moreover, studying land law as an institution in the wider context enables one to draw comparative perspectives on its practicality underpinning different legal systems and thus provide a space for a cross-cultural critique on the value-laden conceptions, principles and practice of the law in different locations. This will help develop a systematic, reflexive and self-critical approach to the study of land law across different countries (see Cotterell, 1992).

The development of appropriate land institutions is inextricably linked with the overall historical, social, economic and political development conditions. An understanding of land institutional change provides further insights into the actual implementation of land policies and laws. Policy implementation is a process that must evolve, and people have to be enabled to participate in this process because they have the 'know-how' (Pressman and Wildavsky, 1973). The study of institutional change from this angle will enable one to better understand policy implementation and legal practice as a culturally contested and socially constructed process. It is in this process that the values and perspectives of the social and political actors can potentially exert influences on the effectiveness of these institutional measures.

Alternative Land Tenure Models and Sustainable Rural Development

As the conventional wisdoms hold that land tenure security is the key to economic development, one may wonder whose land tenure security it is meant for, what it contains and how it can be attained. Land tenure may carry different meanings for different interpreters, especially community and state. Land as a property should be first interpreted as a set of rules and responsibilities. As Dekker (2001: 15) defines, 'land tenure is the institutional arrangement of rules, principles, procedures and practices, whereby a society defines control over, access to, management of, exploitation of, and use of means of existence and production'. This interpretation further implies that it is a sanctioned social relationship between people – not between people and land itself. This relationship is latent in the daily power struggles for legitimate authority to control, allocate and exploit the land (Benda-Beckmann, 2006). Thus, it is always hard to define exactly what tenure security means for different actors in different contexts and what land reforms do in practice under what conditions on the ground in general (Sikor and Müller, 2009).

There is a need to look into the wider underpinnings of land tenure – poverty and power and focus more on the local processes of their interactions, simply because of the failures of approaches led by state, market and community. Community-led approaches are assumed to be responsive to variations in local institutions and practices, but can also cause rising inequalities within the community (Berry, 1993). As a result, a pro-poor approach is used to examine the changing relationships between land, livelihoods and poverty in the context of rural-urban change and to identify the entry points for changes in land policy reform (Daley and Hobley, 2005). As such, one needs to identify the preconditions that need to be established appropriately before any investment in a land management system is made. And secure, flexible and all-inclusive land tenure, whether customary or statutory, provides the best basis for sustainable rural development (Birgegard, 1993). It is a necessary condition for equitable rural development that would otherwise be predominated by elite capture and the stranglehold of the local state.

The complex social relations embedded in land tenure can further complicate the challenges for sustainable land management. Land tenure security does not just stem from individualistic approaches. As strongly argued, 'the notion that only individual Western-style ownership provides enough individual security to promote an economic take off has been substituted by the opposite notion: only communal tenure (in areas where it still holds) provides enough security' (Hoekema, 2000: 51). Although this statement may overestimate the role and function of collective approaches to land tenure security, it provides a useful re-thinking of conventional ways of land management that overlook the dynamics and conditions of land tenure systems. In the Chinese context, obviously both individually- and collectively-based approaches have proved to be ineffective in securing land rights of the poor. The question remains as to what land tenure arrangements work for the poor in a given context, which ultimately challenges the trajectory of land reform in a given community or a nation as a whole. Contrary to the liberal economists' view on simplistic approaches to land tenure reform, it is argued that 'though ideological arguments on the best ways of organizing agriculture continue, no land tenure system can be adjudged best in abstract. Any judgments concerning a particular system must take note of the institutional and technological conditions in the society and the stage at which that society lies in the transformation from an agrarian to an industrial economy. Judgments should also consider what specific groups and individuals in the society are attempting to accomplish' (Dorner and Kanel, 1971: 1).

Given the concerns of societal interests and dynamics, alternative land tenure models have been proposed as a manifestation of the reorganization of social space as a result of rebundling property rights based on values of inclusivity and exclusivity. These models can take the form of conservation land trusts, community land trusts, land banking and so forth to serve different environmental, social, cultural and spiritual goals of the social economy (Blomley, Delaney and Ford, 2001; Pienaar and Brent, 2008). According to Geisler (2010), alternative land tenure models or property regimes present new property relations between the private, non-profit and public sector and redefine socially acceptable uses and communal interests in landownership.

These models may be strategically utilized to combat poverty and respond to social needs. As a result, a given community can employ the hybridization of private, public and voluntary strategies to work together for a more balanced, qualitative community development. The challenge is to understand how these models can be best structured, applied and made to function according to the local conditions. Moreover, these models are to redress the failures of top-down bureaucratic modalities of land tenure reforms to accommodate meanings of land beyond the notions of property for its productive values. And they should be developed so as to be closely responsive to local livelihoods, connections with broader dynamics of authority, interactions with social inequalities and environmental repercussions (Sikor and Müller, 2009: 1312).

In the Chinese case where local communities are not empowered to determine their preferred institutional arrangements for the use of their land, the collective landownership which was assumed to facilitate sound land management practices, has to a large extent failed to ensure land tenure security and enhance more efficient sustainable rural development on the whole. It has not been a genuine institution for community-centred collective action. The government needs to find ways to foster genuine collective action to address many critical issues of rural development and land use in particular.

Effective collective action can solve many issues that cannot be dealt with by policy and legal institutions. Where land tenure is concerned, it can help identify land rights as conditioned by locality, history, changes in resource condition and land use economy, and social relations. And it can respond to changing conditions that affect land use and property rights. Property rights change over time, and the change occurs through social and power relations and negotiations between different groups. This complexity means that collective action provides the means to respond to changing conditions that affect land use and property rights (Meinzen-Dick et al., 2004). Thus, collective action is a prerequisite for pro-poor land tenure, while tenure security is just one element.

The above theoretical constructions, despite their strengths and weaknesses, have one thing in common, that is, the ignorance of how to define the relationship between land tenure, sustainable land use, participatory governance and sustainable rural development. A sole focus on land tenure security for productive values can be tremendously misleading in this sense. It is interwoven with the overall rural development of a given community. Other conditions of rural development have an impact on how land ought to be utilized. Consequently, the applicability and sustainability of a land tenure system is contingent upon many biophysical, social, economic and political factors. Land is just one development sector; other sectors contribute to development as well. The failure to make other sectors work for the poor can also trigger land tenure insecurity and challenge the existing pattern of a particular land tenure system, as the rest of this book will illuminate. Simply, peasants have to decide on whether they should stick to their land in the village or abandon it before migrating to cities. If the village economic and livelihood conditions are not conducive to their continued residence in the village, they would probably go to cities no matter how secure their land tenure is. The outcome would be complex from a rural sustainable development perspective, which further implies the linkages between land tenure and sustainable rural development.

Thus, a pro-poor alternative land tenure system, as this book attempts to articulate and develop, ought to be based on the sustainable land use, rural development and rural governance needs of a given community, whose understanding of local economic and natural resource conditions is inextricably interwoven with social and political relations among different stakeholders. Local community must be given ample power by policy-makers in testing out their preferred choices over a particular land tenure system. A land tenure system can only sustain itself if it contributes to sustainable land use, rural development and governance. A pattern of land use and development further complicates the suitability of a particular land tenure system imposed upon the local community by policy-makers. In essence, the challenge for the design of a pro-poor land tenure system relates to how to match the divergent interests of different stakeholders, especially the poor, for the sake of sustainable land use, equitable development and participatory governance.

Conclusion

This chapter traces the latest developments in relevant theories in land tenure and land reform based on the international context under the domination of the Western economic model vis-à-vis the assumed China model. It is in this context that China's pursuit of the market economy has become complex and controversial, and it remains difficult to chart a clear roadmap for change. By discussing the tenets of failures of land reform programmes led by different institutions, it articulates the need to think beyond the narrow domain of land tenure security as a panacea to a market-oriented scaled rural economy. It proposes alternative models of land tenure that should be more compatible with local conditions. Thus, the dynamics and conditions of land tenure should be studied in a broader context of development to be applicable to the local setting and to be sustainable.

To apply the alternative models, the juxtaposition of households and collective institutions as social, political and economic units can serve as the units of analysis, which will provide an understanding of peasants' experiences, knowledge and relations to power and agency (Croll, 1994). By doing so, according to Campbell (2004), one can more clearly specify the underlying mechanisms for a process in which change can occur; in particular, any constraints and opportunities for change should be carefully examined. It is important to investigate how different actors build and modify the institutions to serve their own interests. The study of land tenure and property rights should provide more in-depth analysis of their interactions with society and the capacity of the state to create appropriate institutions. Land institutions underlie complex social relations that can only be understood through an in-depth investigation into how Chinese rural society is structured and governed and how the meaning of these relations is constructed by the local culture or the perceptions and understandings of the property relations and daily livelihood practices of local communities.

References

Abram, Simone. 1998. "Introduction: Anthropological perspectives on local development", in Simone Abram and Jacqueline Waldren (eds), *Anthropological Perspectives On Local Development: Knowledge and Sentiments in Conflict*, London: Routledge.

Adams, Martin. 2003. "Land tenure policy and practice in Botswana: Governance lessons for southern Africa, *Australian Journal of Development Studies*, XIX (1), 55–74.

Allina-Pisano, J. 2004. "Land reform and the social origins of private farmers in Russia and Ukraine", *Journal of Peasant Studies*, 31 (3 and 4), April/July, 489–514.

Benda-Beckmann, Franz Von. 2006. "The multiple edges of law: Dealing with legal practice in development practice", *World Bank Legal Review: Law, Equity and Development*, Vol. 2.

Berry, S. 1993. *No Condition is Permanent: The Social Dynamics of Agrarian Change in sub-Saharan Africa*, Madison: University of Wisconsin Press.

Biezeveld, Renske. 2002. *Between Individualism and Mutual Help: Social Security and Natural Resources in a Minangkabau Village*, Delft: Eburon.

Birgegard, Lars-Erik. 1993. "Natural resource tenure: A review of issues and experiences with emphasis on sub-Saharan Africa", Swedish University of Agricultural Sciences, *Rural Development Studies*, No. 31.

Blomley, Nicholas, Delaney, David and Ford, Richard T. 2001. *The Legal Geographies Reader: Law, Power and Space*, Oxford: Blackwell Publishers Ltd.

Bromley, Daniel. 2005. "Property rights and land in ex-socialist states: Lessons of transition for China", in Peter Ho (ed.), *Development Dilemmas: Land Reform and Institutional Change in China*, London and New York: Routledge.

Campbell, John L. 2004. *Institutional Change and Globalization*, Princeton: Princeton University Press.

Cashin, Sean M. and McGrath, Gerald. 2006. "Establishing a modern cadastral system within a transition country: Consequences of the Republic of Moldova", *Land Use Policy*, 23, 629–42.

Cotterrell, Roger. 1992. *The Sociology of Law: An Introduction* (2nd edition), London: Butterworths.

Croll, Elizabeth. 1994. *From Heaven to Earth: Images and Experiences of Development in China*, London and New York: Routledge.

Daley, Elizabeth and Hobley, Mary. 2005. *Land: Changing Contexts, Changing Relationships, Changing Rights*, paper commissioned by the Department for International Development (DFID).

Deininger, Klaus, Feder, Gershon, Gordillo, Gustavo et al. 2003. "Land policy to facilitate growth and poverty reduction", *Land Reform, Land Settlement and Cooperatives*, special edition, Washington: World Bank and Rome: FAO.

Dekker, Henri. 2001. *A New Property Regime in Kyrgyzstan: An Investigation into the Links between Land Reform, Food Security and Economic Development*, PhD thesis, University of Amsterdam.

Dorner, Peter and Kanel, Don. 1971. "The economic case for land reform: Employment, income distribution, and productivity", *Land Reform, Land Settlement, and Cooperatives*, 1971, No. 1, FAO.

Ellman, Michael. 2003. "Expectations and reality: Reflections on a decade of agricultural transformation", in Max Spoor (ed.), *Transition, Institutions, and the Rural Poor*, Lanham: Lexington Books.

Gardner, Katy and Lewis, David. 1996. *Anthropology, Development and the Postmodern Challenge*, London and Sterling: Pluto Press.

Geisler, Charles. 2010. (first published online) "Returning land tenure to the forefront of rural sociology", *Rural Sociology*, 58 (4), 529–31.

Grillo, R.D. and Stirrat, R.L. (eds). 1997. *Discourses of Development: Anthropological Perspectives*, Oxford: Berg.

Hann, Chris and Hart, Keith. 2011. *Economic Anthropology*, Cambridge: Polity Press.

Ho, Peter 2003 "Contesting rural spaces: Land disputes and customary tenure in China", in Elizabeth Perry and Mark Selden (eds), *Change, Conflict and Resistance*, London: Routledge.

Ho, Peter (ed.), 2005 *Developmental Dilemmas: Land Reform and Institutional Change in China*, London and New York: Routledge.

Ho, Peter and Spoor, Max. 2006. "Whose land? The political economy of land titling in transitional economies", *Land Use Policy*, 23, 580–87.

Hoekema, André J. 2000. "Keuzevak Rechtssociologie: Rechtspluralisme en ontwikkeling", text on legal pluralism and development, Faculty of Law, University of Amsterdam.

Hutton, Will. 2006. *The Writing on the Wall: Why We Must Embrace China as a Partner or Face It as an Enemy*, New York: Free Press.

Kleijn, W. M., Jordaans, J.P., Krans, H.B. et al. 2006. "Property law", in Jeroen Chorus, Piet-Hein Gerver and Ewoud Hondius (eds), *Introduction to Dutch Law*, 4th revised edition, Alphen aan den Rijn: Lluwer Law International.

Lewis, David. 1999. "Revealing, widening, deepening? A review of the existing and potential contributions of anthropological approaches to 'third sector' research", *Human Organization*, 58 (1), 73–81.

McEwen, Alec and Nolan, Sharna. 2007. "Water management, livestock and the opium economy: Options for land registration", Working Paper Series, Afghanistan Research and Evaluation Unit, http://www.areu.org.af/Uploads/EditionPdfs/701E-Options%20for%20Land%20Registration-WP-print.pdf, accessed 25 February 2008.

Meinzen-Dick, Ruth, Pradhan, Rajendra and Gregorio, Monica Di et al. 2004. "Understanding property rights", in Ruth S. Meinzen-Dick and Monica Di Gregorio (eds), *Collective Action and Property Rights for Sustainable Development*, Washington, DC: IFPRI.

Mertha, Andrew and Zeng, Ka. 2005. "Political institutions, resistance and China's harmonization with international law", *The China Quarterly*, 182, 319–37.

Nonet, Philippe and Selznick, Philip. 1978. *Law and Society in Transition: Towards Responsive Law*, New York: Harper and Row Publishers.

North, Douglass. 1990. *Institutions, Institutional Change and Economic Performance*, Cambridge: Cambridge University Press.

Palacio, Ana. 2006. *Legal Empowerment of the Poor: An Action Agenda for the World Bank*, unpublished draft.

Pienaar, J. and Brent, A.C. 2008. "A model for evaluating the economic feasibility of small-scale biodiesel production systems for on-farm fuel usage", Working paper, NRE, CSIR, Pretoria.

Pospisil, Leopold. 1974. *Anthropology of Law: A Comparative Theory*, New Haven: Yale University Press.

Pressman, Jeffrey. and Wildavsky, Aaron. 1973. *Implementation: How Great Expectations in Washington are Dashed in Oakland*, Berkeley and Los Angeles: University of California Press.

Rutherford, Blair. 2004. "Desired publics, domestic government, and entangled fears: on the anthropology of civil society, farm workers, and white farmers in Zimbabwe", *Cultural Anthropology*, 19 (1), 122–53.

Sikor, Thomas. 2006. "Politics and rural land registration in post-socialist societies: contested titling in villages of northwest Vietnam", *Land Use Policy*, 23, 617–28.

Sikor, Thomas and Müller, D. 2009. "The limits to state-led land reform: An introduction", *World Development*, 37 (8), 1307–16.

Spiertz, Joep and Wiber, Melanie (eds). 1996. *The Role of Law in Natural Resource Management*, Gravenhage: VUGA.

Törhönen, Mika. 2003. "Sustainable land tenure and land registration in developing countries, including a historical comparison with an industrialized country", *Cadastral Systems*, third special edition of *Computers, Environment and Urban Systems 2003*, 1–28.

van Vliet, Lars P.W. 2000. *Transfer of Movables in German, French, English and Dutch Law*, Nijmegen: Ars Aequi Libri.

Wang, Weiguo. 2006. *The Dutch Civil Code* (translated), Beijing: China University of Political Science and Law Press.

Ye, Jianping. 2000. 中国农村土地产权制度研究 (*Study of Rural Land Property Rights in China*), Beijing: China Agricultural Press.

Chapter 3

A Brief Account of 600 Years of China's Land Struggles before the Market Reform Era

Introduction

> A government which permits exploitation of the mass of its fellow citizens … may make a brave show, but it is digging its own grave. A government which grapples boldly with the land question will have little to fear either from foreign imperialism or from domestic disorder. It will have as its ally the confidence and good will of half-a-million villages. (Tawney, 1939, cited in Wong, 1973: xxiv)

Tawney's thought-provoking standpoint on the land question has far-reaching implications for understanding the role of land in China's history, underpinned by social, economic and political inequality and its associated mass struggles for better livelihoods and social and political developments. His inclination to the role of the Chinese Communist Party in successfully instituting land reform to abolish the feudal land-exploitative relations cannot be neglected (see Wong, 1973). In defeating the Nationalist government, the Party-led land reform played an essential role in social and political mobilization characterized by unprecedented land redistribution, which stood in huge contrast to the land reforms undertaken by any previous regimes.

While chapters 1 and 2 briefly highlight the contentious issue of landownership, whose importance to economic development and social stability is unquestionable to mainstream political and scholarly thinking, this chapter tends to bring an historical perspective that underpins a more holistic understanding of its controversy. As China has reached a critical stage of development marked by rising inequality, it is useful to reassess the trajectory of its land reform to understand the lessons learnt from its past. As Perry (2008) contends, few of us now take a renewed look at the past reforms in relation to today's problems.

Land reforms initiated by successive regimes to grapple with peasant rebellions in order to prolong their reigns provide a fertile ground for a reassessment of the nature of these reforms. Their limited successes and large-scale failures had much to do with their inability to address the fundamental issues of social structures and organizations that put the poor peasants on the margins of development. By contrast, the stronghold of local power enmeshed in complex social and political relations was largely reversed by the land revolution led by the Communists at great cost in mobilizing mass support to gain political control in the vast Chinese countryside. However, all these reforms took different shapes to serve more the interests of the state rather than those of the peasantry. This chapter discusses their underlying challenges especially concerning poverty, power and institutions that continue to constrain peasant choice over the reform trajectory. It posits that the creation of genuine peasant-centred land institutions would be indispensable to tackle the needed changes in poverty and inequality that are carried well into the twenty-first century.

It is impossible to provide a detailed account of China's land reform history that spans over 600 years. Rather, this chapter attempts to bring a few thoughts on the nature of land reform and its development underpinnings to the fore. For the first time in Chinese history, the Ming and Qing dynasties saw the sprouts of capitalism whereby land was the crucial issue. In fact, private landownership did exist and land could be transacted even dating back to Ming and Qing and earlier dynasties. This might explain why the Communist-led land revolution, being radical in general, raised the Party's serious concerns about its unintended consequences in the face of mass discontent from the private owners or landlords (Wolf, 1973). This means that land reform has always been inextricably linked with social and political relations structured by the complex vested interests of different actors, which can either facilitate or constrain any reform agenda. In other words, the land revolution aimed at equalizing land distribution might not have achieved the outcomes as envisaged by the reformists, since it was hard to completely dismantle the social fabric of the Chinese countryside.

Starting from this angle it would be interesting to discuss the multi-faceted nature of land relations, peasant struggles and their bearing on the economic and political pressures on the land. The study of land tenure reforms since the Ming and Qing is an attempt to explore the complex relationships and struggles among different actors – landlords, peasants and the state. This chapter concludes that land reforms should not be exaggerated in their effects on the alleviation of rural poverty and equitable development. Rather, it is the inequality between the Chinese peasantry and the dominant rural land elites and the state that perpetuates the institutional constraints to the formation of more meaningful peasant-centred land reform programmes.

Land Reform in the Ming (1368–1644) and Qing (1644–1911) Dynasties

China has been predominantly agrarian, especially prior to 1949 when the People's Republic was founded. Land reforms in China dating back to a few centuries prior to the Ming Dynasty explain a critical fact that land and labour had been extremely important to peasant livelihoods and the rural economy. Since the Ming, the land–labour ratio tended to decrease dramatically, which indicates that with population increase, land had become a scarce resource for the peasantry. Smallholders constituted a considerably large peasant group that dominated agriculture. This group owned land through inheritance or exploitation of idle land. The rural population grew from 65–80 million at the beginning of the Ming to 540 million by the middle of the twentieth century (Perkins, 1969: 16, cited in Lardy, 1983). In particular, by the end of the Ming, the per capita acreage of farmland had dropped substantially, and reached a new record low in the nineteenth century. The incompatible land and labour relations can be seen as a major stumbling block to livelihood betterment and thus a latent factor for land policy changes throughout Chinese history.

Increased population pressures on the land caused its shortage and fragmentation. The latter was partly caused by land sales. At the beginning of the eighteenth century, royal and government land accounted for 27 per cent of the total land, temple land 14 per cent, military colonization land 9 per cent, and the rest with the private holders – individuals or clan corporations. Most peasants had access to the land either through inheritance or through a complex set of leases and rents (Wolf, 1973: 106). Due to serious shortages of land, demand was much stronger than supply, resulting in a sellers' market in China. In the Ming and Qing, land fragmentation became severer than in the past (Chao, 1986). It appeared that it mattered more to the poor smallholders and tenants than the landlords who managed to enlarge their land holdings through amalgamating the land of the smallholders. The latter, in many cases, had to give up their land due to various economic pressures including tax payments. As a result, land fragmentation for the poor and land concentration in the hands of the mighty few, including the landlords, appear to be a major factor for peasant–landlord or peasant–state struggles. In addition, internal rifts among the gentry, rich and poor peasants invoked by the changing economic conditions, for instance, the injection of capital into the countryside, was another major trigger for the vicious cycle of peasant rebellions and social and geo-political disintegration followed by new waves of consolidation and integration (Wolf, 1973). Moreover, natural disasters constituted another major factor for peasant rebellions. From 1626 till the end of the Ming, for instance, famine, drought and other calamities befell northern China, which severely curtailed agriculture. This was exacerbated by the government's inability to collect taxes, which even led to troop rebellions that severely contributed to the demise of the Ming dynasty.

Changing Land Relations

In the Ming and Qing, there were diverse types of landownership patterns. Land could be owned by landlords, smallholders, soldiers and gentry. The land of the gentry was allocated by the emperor and increasingly turned into private land. The landlord group consisted of empire officials or gentry and ordinary peasant landlords who had less political power and social status than the former (Lin and Chen, 1995; Li, 2007). The outline of the rural social structure underpinned by land tenure in late imperial China is provided in Table 3.1.

Table 3.1   **Outline of rural social structure underpinned by land tenure in late Imperial China**

Gentry: An important state group in the scholar-official category. Refers to those who had qualified for office in the imperial bureaucracy by passing state examinations. They formed the core of the local elite in each district. They were often absentee landlords, but not all were large landowners. By the end of the nineteenth century, together with their families, they comprised an estimated 7.5 million people or 2 per cent of the total population.

Managerial landlords: Employed three or more long-term labourers and directly managed part of their estate and sold part of the surplus product for profit. Many of them rented out parts of their estate. Also engaged in rural business enterprises, and most of them lent money at high interest rates.

Rentier landlords vs leasehold: Did not manage their estate directly but rented out to tenants. Leasehold system allows the tenants to have full land use rights in terms of transfer, inheritance and sales.

Rich peasants: Employed fewer wage labourers and farmed less land than the managerial landlords; engaged in commerce, handicrafts and usury.

Long-term labourers: Employed by the landlords to work from one month up to one year. They frequently owned small plots of land, but sometimes owned nothing at all. Wages were their main income – half paid in cash, the other half in the form of meals.

Short-term labourers: Employed during the busy seasons in special rural labour markets, usually in the local market town. They owned small plots of land, and income was derived from their land and wages as well as from secondary occupations such as being peddlers, stone-cutters, mat-weavers, etc.

Source: Author's own compilation, based on Wilkinson (1978), pp. 11–13, 23–4; Wolf (1973).

The rural social fabric shows complex land relations. The landlords, especially those with close links with the empire, had more economic and political privileges over the others as they were levied fewer taxes and required to contribute little labour to the state. In many cases, these burdens were transferred to the poor peasants, many of whom were forced to give up their land to the landlords in order to avoid the heavy taxes imposed on them. In case of being indebted to the landlord in terms of unpaid loans, they were more likely to become tenants. During the mid-Ming period, the landlord managed to profit from amalgamating the land of poor peasants, which had actually adversely affected land sales, although land sales had started on a small scale. Those smallholders were also affected as they could not sustainably maintain their land and property.

According to Chao (1986), with continuous population growth, more and more smallholders became tenants in view of scarce economic opportunities. The tenants' rights and social status gradually gained legal recognition. As a result, the leasehold system became less favourable to the tenants, whereby the landlords increasingly lost strong control over their tenants who had gained more freedom to move in and out of the land and thus some became relatively free hired labourers. To a certain extent, the freedom tenants gained was conducive for the sprouting of capitalism because of the possibility for them to invest in land for their own interests. Being wage labourers meant more savings for their own investments (Li, 2007). With the development of rural-urban markets, many Ming peasants did not rely on subsistence farming any longer; instead, they produced products for the market for profits (Heijdra, 1998).

Land reforms and economic development had a profound impact on the changing land relations. Collisions between land policy changes and traditional or customary land relations occurred and sometimes rural communities' resistance to change was prominent. The Qing saw the customary land laws resting on kinship relations as barriers to the development of land markets. Since under customary laws land transfers were given preferences to priority parties – relatives, neighbours and other close affiliates over third parties – the latter were only allowed to buy the land not wanted by the other two groups. This was seen as an impediment to the formation of a land market and a major contributor to land conflicts. As land sales increased, the abolition of this system was put on the empire's reform agenda with a view to safeguarding the interests of sellers to ensure that they could sell the land at favourable prices. Hence, this agenda was seen as a move to make way for free land trading (Li, 2007). As traditional households or lineage-based groups became larger and larger, more and more divisions and resentments over inequitable distribution of land-related benefits further undermined the functionalities of these institutions in community unity and development. Gradually, in many localities, these customary institutions had begun to be phased out (Menzies, 1994).

Land commercialization was firmly entrenched from the late 1700s as was evident in the inter-regional land sales. In the 1840s, it was additionally increased by the merchants who issued high-interest loans to poor peasants. In times of insolvency, the poor peasants had no choice but to give up their land; the incentives of the merchants to do so also lay in land-induced investments in agriculture. This phenomenon did not indicate that individual peasants were worse off. In fact, many peasants became rich, and so did the landlords, while some landlords even joined the gentry group. With the advent of commercialization of foodstuffs and the development of cash crops, further group divisions among the peasants took place. Some poor smallholders became tenants, while some rich peasants and a minority of smallholders gained more development opportunities. However, during the Qing, the landlordism associated with the empire had not retreated, although it had less power than in the Ming in terms of its ability to seize land from the peasants and avoid land taxes. Extensive land was often accumulated in their hands and became a major source of their wealth. In addition, they gained lucrative profits through high-interest loan schemes. Although their relationships with their tenants were not as harsh as those found in the Ming, the landlords maintained political and economic privileges in the local polity (Li, 2007).

Since the early Qing, changes in agricultural production relationships between landlords and tenants took place. More and more tenants had become wage labourers on the farm. The landlords had begun to realize that it would be more profitable to hire labour directly for agricultural production than renting their land to the tenants. With the growth in population and improved labour availability, the tenants did not have to tie themselves to the land. At the same time, many gentry and landlords associated with the rulers would only consider expanding their land area to increase their rents rather than improving agricultural efficiency and harvests. Many of them became absentee landlords as they left the countryside for the cities. In cases where written contracts between the absentee landlords and the tenants provided tenure security for the tenants, the latter stood a chance of becoming the owner of the land he worked and could sell or mortgage it at will (Beattie, 1979).

In sum, the changed land relations in the Ming and Qing reflected the fact that the land tenure system gradually shifted from the predominance of the feudal landlords to an increased role of the peasantry over landownership and use through tenancy and wage labour. As a result, agriculture had developed and contributed to economic prosperity, especially during the early Qing. Of course, this progress was also due to the improved policies of the regimes to offer more incentives in land use to the peasantry. However, they did not actually address the fundamental issue of landownership. Although the majority of the peasants had gained more political freedom and land rights than ever before, land was still controlled by the imperial state and landlordism. As the Qing regime represented the minority 'Man' ethnic group, its policy measures to weaken the majority 'Han' were seen as a strategy to restore order and strengthen their own control systems. Thus, it had contributed to the restoration and strengthening of the embedded feudal relations.

Moreover, the nature of landlordism and the stronghold of feudalist productive relations disarrayed the development of land markets and the agricultural economy. Although the overall trajectory of capitalist production was inevitable and even grew stronger, feudalist landlordism was continuously reproduced, which further slowed down agricultural development. The other important factor was the role of the state in safeguarding the interests of landlords and their power over the peasantry. As a result, China did not enter the normal phase of agricultural capitalism. As earlier mentioned, the reinstatement of the power of the landlords, gentry and the imperial state over the Chinese peasantry after peasant struggles further verified the stranglehold of the rural social structure and relations in controlling the masses.

The Nexus between Peasant Struggles and Land Reforms

To contain rising conflicts between landlordism and the peasantry, the empire, as seen in all dynasties, tried to adjust land taxes and other related obligations of the landlords to undermine their power in order to reduce peasants' land-induced economic and physical burdens and thus to improve their land rights. This made it difficult for the landlords to acquire more land and invest in it. In the Qing, the landed gentry cared little about how they could maximize land production. Moreover, the widespread nature of small landholding made the peasants extremely vulnerable to land-induced natural and economic risks, which was a major factor in frequent peasant protests in Chinese history.

Despite the strengthened role of the peasantry in land tenure developments, one should not underestimate the harsh land relations and struggles and take for granted that economic measures could do justice for the disadvantaged. The Chinese social structures underpinned by the predominance of the state and its associated landlordism and gentry over the masses can be seen as a major factor in peasant struggles. The trajectory of land tenure reform from the founding of the Chinese empire right through the demise of the Qing clearly shows that social inequality and injustice between the two major groups posed a threat to social and political stability and economic development, which calls into question efficient and sustainable land use.

This argument contradicts the claims that land tax systems in both dynasties were exploitive forces against the peasantry, which constituted the primary causes for peasant rebellions. In fact, China has long been an agrarian society marked by intensive farming carried out largely by the peasantry. As the population grew, it was extremely difficult for the peasants to feed themselves on their tiny plots of land. As a result, they had to expand their cultivated land and enhance grain yields substantially in order to meet their basic needs (Wang, 1973: 6–8). As the expansion of land acreage reached a limit, land struggles for subsistence needs became more pressing in rural China. It is not difficult to understand that land as a scarce resource thus could be conducive to peasant demands for better livelihoods and equity.

It is important to note that the Chinese imperial regimes' land tenure reforms were partly to compromise peasant appeals for equal land redistribution and exemption of varied duties and partly to maintain their power and control over the local landlordism and peasantry. Strikingly, all these reforms had one thing in common – redressing social and economic inequality through the so-called egalitarian principles and methods, which were highlighted by streamlining the tax and labour obligations of the tenants and small-scale land redistribution while disadvantaging landlordism. To a certain extent, these measures were useful in curtailing the exploitive power of the landlords. During Ming rule, for instance, the 'One Whip Law' was a major instrument in synergizing varied taxes and obligations and converting them to land-based obligations to the empire. As a result, peasants gained more freedom of choice in land investment and business activities (Chinese History Textbook Net, 2009).

However, the sustainability of these reforms became problematic in terms of their effects on equity and poverty reduction for the disadvantaged groups, which explain the subsequent peasant rebellions. At the aftermath of each reform, for instance, it was not uncommon that the early Qing dynasty promulgated relevant measures to redistribute some lands to the peasants with the aim of ensuring tax collection and consolidating its control in the countryside. It even issued land certificates to those who were allocated land. In addition, the land forcefully taken by landlords was returned to the original owners who were obliged to pay land taxes, and their landownership was recognized by law. In order to guarantee income from tax collection, among other reasons, the empire did not have the intention, or it was simply impossible, to abolish landlordism. Instead, it recognized the legal privileges of landlords, while punishing and restricting their illegal behaviour. Moreover, the regime was directly involved in enclosures of the land under their direct jurisdiction and forged a new privileged landlordism. As a result, the basic social structure and the power of social and political groups remained largely intact.

When peasant struggles resurfaced in 1712, the Qing regime promulgated a considerably more relaxed rural taxation system (*tan ding ru mu*) aimed at reducing the taxes based on the number of household members. In spite of an increase in the number of household members, the household was no longer required to pay more taxes under this new policy. The latter led to a combined land and labour taxation system that stipulated the levying of land tax that subsumed poll tax and labour corvée. Consequently, poor households gained more freedom and time to spend on other economic activities. Additionally, it put more pressure on the landlords to pay land taxes, which set limits to land concentration and thus eased social tensions. In a nutshell, it was a further improvement to the 'One Whip Law' implemented during the mid-Ming period. It is also noted that the two systems resulted in the inability of the empires to collect sufficient taxes from the landlords and the peasants whose struggles had a negative impact on the national economy and land utilization.

Nevertheless, the limited success of these reforms did not trigger rapid development of more equitable land relations. Prior to the demise of the dynasties, land became re-concentrated in the hands of the mighty few, whose exploitation of the tenants further deepened their conflicts, obstructed the development of a market economy and, moreover, weakened the state's control of the local landlords. To a certain extent, the massive protests and the deposition of the Ming were coupled with agricultural development as demonstrated by the increases in food production and commercialization. Land sales involved business investors who hardly existed in the past. In times of aggravated poverty and natural disasters, these groups provided high-interest loans to the peasants who mortgaged their land and had to sell it at lower prices when they could not repay the loans. Some of the investors eventually became new tenants. As a result, land titles frequently changed hands, and ownership gradually concentrated among the big buyers, who at times used force to obtain the land (Li, 2007).

The Qing dynasty also saw the re-accumulation of land by the landlords. The latter, especially represented by the royal family members and the gentry, turned many peasants into their tenants in the mid-nineteenth century. Once again, the high tributes paid to the landlords by these tenants caused deep discontent with and hatred for their masters. The peasants desired a better life based on equalization of land rights distribution and its associated economic obligations within the entire social stratum. As a result, it was also during this period that the empire encountered the harshest peasant protest known as the Heavenly Kingdom Revolutionary Movement or the Taiping Rebellion that affected roughly a third of Chinese territory over a decade at the cost of a death toll of about 20 million (1850–64). This movement promulgated the most comprehensive land reform agenda in Chinese history – the Heavenly Kingdom Land Law featuring land equity for the peasantry, including women through land redistribution, in 1853. In this law, agriculture was to be organized around units of public and private farms cultivated by the peasants. Moreover, it envisioned a new social order against the rule of the Chinese gentry and their ideology and Confucianism. Thus, the Taiping Rebellion is regarded as the forerunner of modern social movements. However, the rebellion itself paid less attention to improving the lot of the peasantry than organizing it to suit the needs of the new social order, in which the peasants would remain as the main burden-bearers of the envisaged society. As a result, the agrarian reform programme was not realized, for it could only count on the loyalty of the peasants to a limited extent (Michael, 1966). Despite the failure of the uprising, which was brutally suppressed, the recognition of the system's resistance to the feudalist land relations was far-reaching. However, it also received considerable criticism for its idealistic and unrealistic approach to development and social justice (Chinese History Textbook Net, 2009).

The preceding account of land relations, peasant struggles and institutional reforms also reveals a crucial fact – limited land concentration and large social fragmentation in the two dynasties. Households and their descent groups as social organizations developed their own rules governing land use embedded in the institution of clans and provided support and protection for their members. It was difficult for large landowners to consolidate their estates over many generations. Land concentration was actually a slow and hazardous process. For instance, to accumulate a few hundred *mu* of land could take a household no less than one hundred or a few hundred years over many generations. And even big landowners in the process could return to the status of small owner-peasants in the face of a rapid succession of household divisions coupled with poor land management. By the end of the nineteenth century, few landlords owned more than 10,000 *mu* (1,700 acres) (Wilkinson, 1978: 17; Menzies, 1994). This relatively low level of land accumulation was a further indication of the nature of land fragmentation and small-scale farming in general in the Chinese countryside. For the imperial state, it was easier to impose taxes on the smallholders than the gentry. The formation of large landholdings had always been seen as a potential challenge to state domination in the Chinese countryside (Huang, 1985).

In a nutshell, land reforms were implemented by each regime in a cycle of reinforcement instead of separate and irrelevant initiatives. It is difficult to delineate land relations due to the lack of systematic historical analysis. The vortex of enmeshed struggles and reforms explains the failure of the two dynasties to adjust imbalanced land relations through land distribution and taxation reform, among other measures, primarily due to their inclination to economic measures rather than social and political mobilization of the peasantry.

Land Relations as Class Struggles?

For some historians, especially those with the inclination to the Marxist view on economic development, unequal social relations are the primary causes of peasant rebellions, which gathered momentum during the Qing. This claim is based on the identification of some primary landlords during that period (Chao, 1986). However, others disagree. The ruling class was claimed to be represented by the Chinese gentry, which was a tiny and mobile group. This group gained its wealth and influence entirely through its possession of formal educational qualifications and office, thus their status was not based on the holding of large landed estates (Beattie, 1979). As Chiang (1982) contends, according to the records of land registration in several localities,[1] most land was owned by peasants, and since the mid-Ming, there had been a trend towards diversification of landownership. This finding is further verified by Wilkinson (1978: 9), whose calculation of 200 villages in Shandong Province in 1900 shows that owner-peasants counted for 55 per cent of the surveyed population, followed by 17.1 per cent wage labourers, 16.9 per cent tenants, 4.6 per cent rich peasants, and only 3.8 per cent landlords.

Land taxes, poll tax and labour corvée services were levied on the smallholders whose incomes from their tiny land plots made it extremely hard to meet the government's demands. Instead, they would rather seek tax shelter from other large landowners even by donating their lands to the latter. But as already discussed, land taxation reforms had a major impact on land concentration, which was less important for land distribution after the fifteenth century. Growing commercial activities in urban China and low economic returns in farm production had caused many wealthy landowners to become less interested in land investments. Many large landlords left their homes for cities and left their lands to smallholders. By the early nineteenth century, the Chinese countryside had become dominated by smallholders – peasant owners and petty landlords, who owned slightly more land than the well-off peasants (Elvin, 1973).

Coupled with the exodus of the landlords to the cities was the injection of capital into the countryside by the merchants who invested in land purchased from those smallholders who failed to earn enough income from the land. Many argue that this phenomenon led to the concentration of land in the hands of the merchants. However, some hold that the growth of commerce virtually caused many disincentives in landownership among merchants. There was also a diversion of capital from rural land markets to the urban sector (Elvin, 1973). There is a lack

---

1   Farmland registration was required in the Qing dynasty. Each landowner had to register his land with the local government for tax assessment and ownership identification. A registration serial number was assigned to each plot and a survey map was included. See Chao (1986).

of consensus on whether this trend would have facilitated land concentration. But given the nature of land fragmentation and the degree of land smallholdings, it can be seen that both commercialization and the inflow of capital into the rural areas tended to disperse what would otherwise have been large-scale land concentration. Farmland in traditional China was gradually owned by increasing numbers of small- and medium-sized holders. Furthermore, as land was so dispersed and fragmented, direct investment might not be too conducive for the landlords, some of whom would rather reserve plots of land for their own purposes and lease the rest to others. As a result, tenant farming was prevalent in the sixteenth century (Chao, 1986: 117).

Over a thousand years or so, the demise of extensive landownership, without much influence from various land equalization policies imposed by successive regimes, calls into question the class issue. Those who were called landlords in the early twentieth century in the Republic era were small- and medium-sized holders. Peasant rebellions were never the exclusive work of any one social class or group in Chinese society because the peasants were not the only actors in the movements. Other groups like the intelligentsia played a key role in organizing the struggles. Although they might represent peasant interests, most of the movements were not initiated by the peasants themselves. Rather, the organizers could be from other social groups, who had broader interests than the immediate needs of the peasantry. Thus, it is simplistic to view these movements as articulate class struggles. According to Hsiao (1967: 511), ordinary Chinese peasants have just one dominant desire – sufficient means to keep their families alive by possessing a piece of land and all that it yields. Whenever a movement promised them better livelihoods through land redistribution, they followed. Hsiao further contends that the concept of peasant revolution may be useful or indispensable to propaganda purposes but it can hardly withstand objective historical analysis (also see Michael, 1966; Wolf, 1973). Due to a lack of solid and consistent data, it is hard to ascertain the extent to which Chinese society and peasant revolutions were structured on class relations.

Furthermore, Tawney (1966) cautions against the use of the term 'class'. As he points out, the Chinese history of peasant riots was not the consequence of the so-called mal-distribution of landed property. China did not have a powerful landed aristocracy with *de facto* control over the lion's share of the land; nor was there a huge landless peasantry. Rather, the basis of their contradictions and conflicts was the fact that the peasants had nothing more than tiny land plots to cultivate for their livelihoods. In this sense, they should properly be called 'propertied proletariat'. Their impoverishment was further complicated by rising population, a lack of alternative opportunities and the exploitation of the landlords, usurers and speculators as well as the state. All these factors contributed to peasant rebellions.

However, Fei (1980) argues that the concept of class is relevant with the gentry and peasantry constituting two distinct classes. The gentry (20 per cent of the population) was maintained by owning land and having political access to officialdom. Mobility between the two classes was rather limited, as the existence of kinship groups among the gentry provided mutual security and protection (also see Ho, 1962). Irrespective of whether this finding is reliable, it can be seen that the formation of a local elite group with the exclusion of poor peasants had a profound impact on livelihoods and land relations in particular. Land was perceived by the local elite as an essential form of security for wealth accumulation and social mobility (Beattie, 1979). As the rural population was stratified into diverse groups with unequal political, social and economic status, understanding the discrepancies of interests between the rulers and the masses and thus tackling these fundamental rural development problems were not dealt with effectively by the Qing. Failure to do so had contributed partially to its demise (Hsiao, 1967).

It can be seen that the argumentations on land relations as class struggles tend to pay overt attention to the size of land holdings by different groups rather than the challenges of improvement of peasant livelihoods as constrained by the cohort of other social, economic, political and biophysical factors. The size of land holdings cannot fully reflect the extent of land-related social and economic inequality and the multi-dimensions of rural poverty. These inequalities did not stem solely from land relations or exploitative landlordism, but more from the broader social and economic challenges that the successive imperial regimes found it difficult to grapple with. In other words, it is incorrect to claim that inequality is unrelated to land relations, but it is more important to understand the underlying causes of inequality. This explains the failures of the Ming and Qing land reforms in tackling inequality by mainly using economic measures, for instance, through taxation reforms, to ease peasants' burdens induced from the land without addressing adequately other dimensions of land struggles.

Failure of Land Reform in the Nationalist Republic Era (1911–1949)

In the aftermath of the demise of the Qing dynasty, brought about by the Chinese Revolutionary Army in 1911 and followed by the election of Sun Yat-Sen as the first President of the Republic of China, rural China was in a state of destitution. Agricultural development was in a stalemate due to a lack of capital and the overt reliance on traditional farming techniques. For instance, in 1918, about 50 per cent of the peasants were occupying owners, 30 per cent were tenants and 20 per cent owned part of their land while renting the rest. Land fragmentation and small-scale farming continued. Farming was virtually a kind of gardening (Tawney, 1966: 34, 46). Those wealthy landowners had primary interests in rent collection and tax evasion. When they had extra capital, they used it to acquire more land rather than spending it to improve farm conditions and the livelihoods of the tenants (Hsiao, 1967). Population pressures further deteriorated the stagnated agriculture, which was unfit to feed the burgeoning population. The Nationalist government realized these problems and passed measures for the creation of agricultural banks, credit societies and other cooperative organizations. These institutions were deemed necessary to help the peasantry, but they did not succeed due to various economic and political constraints (Tawney, 1966).

By the 1930s the extent of land concentration, tenancy and rural poverty was more severe than several decades earlier. Private landownership was the dominant feature of land tenure and inequality in landownership was prominent. It is estimated that 70 per cent of the households owned less than 15 *mu*, which constituted less than 30 per cent of the cultivated land. The households that owned more than 50 *mu* only accounted for 5 per cent, which is 34 per cent of the cultivated land. Only 1.75 per cent of the cultivated land was owned by households with 1000 *mu* or more. This finding is shared by others. For instance, in 1936 in north China, landlords who constituted 3 to 4 per cent of the population actually owned 20–30 per cent of the land. By contrast, poor peasants formed 60 to 70 per cent of the population and owned less than 20 to 30 per cent of the land. This inequality was magnified in south China (Wolf, 1973: 134). Hence, small peasant farming constituted the overall rural economy characterized by a low level of labour productivity and agricultural technology and declining farm size (Feuerwerker, 1983; Riskin, 1987). Large regional variations in landownership distribution and tenancy were also seen as a stumbling block to balanced development. Central and southern China had more prevalent land concentration and tenancy than the north. In particular, a high-level of land tenancy was mostly found in the south – the Lower Yangzi and Pearl River Delta, which had the most fertile land and commercial areas (Riskin, 1987: 26–29). Many rich peasants obtained their land by renting from others and then subletting it. However, as land tenancy often involved money lending, the poorer tenants had to pay very high interest rates. As banks and credit institutions were scarce in the countryside, their income from the land became insufficient to sustain their livelihoods as only the minority managed (Douw, 1991). In addition, in central and southern China the number of absentee landlords grew. They lived in rural townships or in district towns and left their land to those bursaries who managed the land for them. The absentee landlords charged their tenants fixed rents and showed little care for their tenants' livelihoods (Eastman, 1988). All these factors reflected social inequality in the Chinese countryside and had different effects on different social groups (Wolf, 1973; Huang, 1985).

The uncertain and complex rural land relations were also complicated by the rising tensions between the state and rural society. Heavy state taxes imposed on the rural landowners were seen as a means of state power penetration into the countryside and a cause of rural rebellion (Huang, 1985; Bianco, 1986). As a result, the central government had gradually lost effective control over the countryside, which was more in the hands of the gentry and warlords, who represented an administrative force that undermined the power of the central government. The latter had to depend on the gentry and warlords more than in the past in order to contain social unrest and maintain peace and order; the centre spent increased amount of land taxes locally in order to appease them and cover the costs of mounting local administration (Douw, 1991).

In the 1930s, the Chinese rural economy was hit by the world economic crisis coupled with its inherent constraints, which caused massive rural poverty and unbalanced rural–urban development. Unemployment in the urban sector denied rural labourers any prospects. The impoverishment of the peasantry was also exacerbated by the effects of natural disasters, increased banditry, harassment by warlord troops and the exploitation of the state in exacting taxes. Many poor peasants sold their land, which became a latent and complex cause of land redistribution at a later stage (Eastman, 1988).

In sum, the Nationalist government faced daunting challenges of poverty and inequality interwoven with other land-related issues. It struggled to find a viable solution to the roots of these social illnesses that had been carried over from previous imperial regimes in order to avoid the likelihood of China's path leading to entrapped capitalism, exemplified by the USA or the highly concentrated landownership seen in the UK (Schiffrin, 1957). President Sun was even approached by Lenin for political cooperation shortly after World War I. Their cooperation led to the judgement that communism was unsuitable for China. To Sun's disappointment, the principle of 'equalization of land use rights' and building a free democratic China came to an end as a result of his death in 1925 and the war with Japan. The latter brought the government administration to a standstill. Some measures were taken to lessen the burden of farm tenants who were forced to pay excessive rents to their landlords, far from realizing Sun's reform agenda. Research has revealed that his agenda, focused on land value taxation rather than more aggressive means, had a strong influence on the Communist Party's land reform agenda (Wu, 1955).

Revolutionary and Early Stages of Land Reform by the People's Republic (1921–1978)

It was under the Communist Party led by Mao Zedong that radical land revolutionary reform activities took place in the areas they controlled, characterized by forceful confiscation of land and redistribution among the landless. It is important to note that Mao's ideas on land equity had no major difference from Sun's and those of ancient regimes, for Mao himself even remarked that it was the ideology of all revolutionary democrats and that it was not solely owned by the Communist Party. However, Mao's land revolution carried its own implications for social and political movements in China.

Since its inception in 1921, the Communist Party had set its goal of reforming Chinese society, attaching great importance to uniting and organizing the peasants through the revolution. This was reflected in the well-known strategy of 'encircling the cities from the rural areas'. In 1926, the party-led peasant movement started in Guangdong and quickly spread into Hunan, Hubei and Jiangxi provinces. Its initial mandate was changed from reducing land rents to more rigorously addressing the root of China's land problems. In the centre of this movement – Hunan province – overthrowing landlordism was put on top of the agenda. Land reform was recognized as the key to restoring social order which was in disarray as a consequence of war and conflicts between the Nationalist and Communist parties and chronic poverty. Peasant associations and armies were organized and engaged in all activities targeted at landlords. Many of the landlords including small landlords and petty bourgeoisie were severely harmed or killed. The Communist Party, however, criticized the peasants for taking the law into their own hands and causing violence to these groups and other innocent farmers. But later on, this criticism was reversed by Mao who saw violence as the only way for the success of the revolution. In 1927, the Party claimed that it had a totally different ideology from the Nationalists and put the representation of the interests of the poor – peasants and industrial workers – on top of its political agenda.

During 1927–1937, in general the movement reached hiatus in terms of the use of harsher punishment for landlords, which led to complete confiscation of their land and its conversion into state property through land nationalization. Only during 1930–1931, did the land policy see a minor change in terms of lifting the ban on land transfers to allow for land leasing based on the formalization of the peasant land titles. A major motive for this move was to settle the peasants' concerns over land tenure insecurity caused by radical land expropriations. Yet this policy was short-lived, as from 1931 the violence against the landlords took place on a larger scale with their houses and land deeds burnt to ashes, and many lives were lost. At the same time, the movement against the bourgeoisie, rich and middle peasants was launched. This was in close connection with land violence, as land privatization was targeted to stop land transfers and hiring of labour. It called for the establishment of collective landownership and production in order to implement the most comprehensive socialist policy (Gao, 2007).

When the war against Japan broke out after the Long March, the Party reassessed its land reform process and acknowledged that it was an incomplete success given its cruelty against all groups of the peasantry and the severe damage done to the Chinese countryside.[2] It decided to change its policy from forceful deposition of the Nationalist regime into joining forces with them to defend the country against the Japanese invasion. Subsequently, the land policy shifted from confiscation of landlords' land to reduction of land rents and taxes. As a basic agricultural policy, the policy of reducing land rents and taxes policy was actually written into the 1930 Land Law of the Republic of China, but it was the Communists who managed to implement it. Nonetheless, the implementation of this policy quickly fell into a vicious circle of violence. Widespread violent acts against the landlords, rich and middle peasants resurged. These groups were severely punished with some of their

---

2   The Communist Party had for a long time been under the influence of the Soviets. The Long March enabled the Chinese Communists to free themselves from Soviet influence to a certain extent. As a result, the Party started to rethink the goal and strategy of the revolution, which they felt ought to be country-specific rather than being dictated by the Soviets.

lands confiscated. This situation worsened after the defeat of Japan in 1945 and the revival of new conflicts between the two political parties. This time, any use of violence to take land from the landlords was even encouraged and the protection of their landed property rights became non-existent (Gao, 2007).

As Hinton (1983) clearly pointed out, based on his fieldwork in the regions used as revolutionary bases, the land reform movement had the sole purpose of stopping any possibility for the Nationalist Party to form an alliance with the landlords and aristocracy. The conflicts were so harsh that the peasants at all income levels were afraid of physical and mental abuse. Facing chronic poverty, the poorer seemed to show their dissatisfaction with the movement. This led to the Party's decision to assess the effectiveness of the land reform. Nevertheless, in 1947, the Land Conference of the Communist Party passed the Outline Land Law of China that highlighted the need to emphasize equal redistribution of land to win the civil war, for this was deemed as necessary to meet the demands of the poor and thus to organize them in the combat; for instance, some of this law's mandates are shown in Table 3.2.

Table 3.2   1947 Outline Land Law of China

Article 1	To abolish the land system based on feudal and semi-feudal exploitation, and to realize the land system of 'land to the tillers'
Article 2	To abolish the landownership rights of all landlords
Article 3	To abolish the landownership rights of all ancestral spirits, temples, monasteries, schools, institutions and organizations
Article 4	To cancel all debts in the countryside incurred prior to the reform

Source: Author's compilation based on Wong (1973), 282.

Subsequently, land was once again meant to be redistributed among many peasants. Yet, land redistribution during the civil war was not full-scale equal distribution. In fact, it proved to be a partial reshuffle of agricultural resources – a mere 40 per cent or so of the land was involved in redistribution. Furthermore, confiscated resources were not equally but differentially distributed among the beneficiaries who constituted approximately half the rural population, which means that the redistribution agenda had to compromise with the political and economic reality in order to avoid radicalization tendencies. It is important to note that the reform movement encountered huge difficulties in mobilizing the peasants especially in south China, where there was a high rate of land tenancy. As conventionally conceived, there should be a causal relationship between land tenancy and rural unrest, which proved to be the opposite case in that region. The reasons lie in the fact that successive reforms since the Ming and Qing dynasties had done little to alter the structure of local power embedded in the hands of local gentry, local bandits and their associates – all tied together in close clan relations. In this context, it was extremely difficult for the peasantry to play an independent political role (Michael, 1966). Thus, this reform met obstacles in balancing the political and social costs of land expropriation and the requirements of redistribution for economic efficiency (Wong, 1973). This also explains how the revolutionary strategy of the Communists went through several distinct phases from being radical to mild land reforms aimed at winning the support of the middle and rich peasantry.

Compared with the victory of the massive Communist-led land movement throughout the country, the Nationalist Party is often claimed to have been ineffective in reforming the Chinese countryside. Yet, according to Gao (2007), the judgment should not be that simple. He sees the traditional social structure and organizations as the main obstacle to the reform that could not be transformed at both bottom and top social and political levels. This means that one had to seek irrational ways to launch the reform, which was exactly what the Communists did. Also it was a process in which they managed to learn from the reform practices and gained renewed support of the masses. Moreover, Hinton (1983) provides an insightful account of the reformed villages by the Party. The Party quickly realized after its radical reform that land distribution itself was not sufficient at all to build firm support among the peasantry. For this reason, the Party managed to establish poor peasant units, based on which peasant associations and village cooperatives were formed to fill in the political vacuum in the countryside. Through these organizations, the Party consolidated its control at the lowest level of society, which makes it the largest political party in the world up to today. The violence used in peasant rebellions mentioned earlier further proved the power of mass organizations to change village society – everyone, even the Party members, had to be brainwashed to gain a place in the process of social transformation.

To a large extent, the revolution reversed the structure of Chinese society at the expense of agricultural productivity. In the beginning of the movement, many people opposed the idea of land redistribution for demographic reasons, for this could lead to further fragmentation of farmland. As a result, it could create inequality between the capable and incapable labourers. For instance, in the northeastern region, in the aftermath of the revolution, land productivity decreased as compared with the past because of three factors. First, landlords and rich peasants were severely affected and lost their land to the poorer peasants who were allocated the land through redistribution, but poor peasant recipients were inexperienced in self-organization and production. Second, as elsewhere, a large number of affected rich and middle peasants lacked the incentives to till the land, because they were afraid of personal abuse and wary of land investment. Third, the reform led to the reduction of the labour force as well as livestock. But the opposition force could not withstand the mainstream political force underpinned by the call for mobilization of the masses, neither was it able to offer alternatives for agricultural development.

After the Community Party took power in 1949, about 700 million *mu* of land were redistributed from landlords to landless peasants and tenants, who totalled more than 300 million. The state then took the surpluses held by landlords for rural social welfare and urban industrial development (Esherick, 1981). As Mao ordered, the key task for the Party after the revolution would be the restoration of social order and development of agricultural production. He called for a peaceful solution to China's land problems by taking a cautious approach in dealing with the landlords in order to stabilize the countryside. He even decided not to touch them and leave them alone for some time. However, in general northeastern regions saw more peaceful land reform than the rest of China. Overall, rich peasants were not brought under effective protection (Gao, 2007). This also implies that land redistribution through the revolution seemed to have dismantled the social fabric, but not with as many effects on the real power structures as one simplistically conceived. This also means that a loss of land for one group does not necessarily lead to power change.

From 1953 onwards, Mao initiated the land collectivization programme that reversed individual landownership and reinforced Party micro-management (Spence, 1999). Under collective management, there was a lack of economic incentives and motivation for the masses as well as for the local bureaucrats, who had no resources to improve agricultural efficiency (Wu and Reynolds, 1988). This situation was exacerbated by the Great Leap Forward 1958–1961. Aimed at boosting economic growth, it created huge centrally managed projects that involved up to 100 million peasants to open farmland, create people's communes and develop industrial capacity. With very limited success, it caused a severe decrease in agricultural output, which led to mass starvation. It is widely claimed that this movement was driven by economic incentives in terms of prioritizing industrial development, and more importantly, the wish to forge a new identity for the Chinese. By doing so, the state exerted more political and ideological control of its subjects. This was seen as a way to keep the Marxist-Leninist doctrine intact in the face of intra-party political struggles which gained momentum during the Cultural Revolution (1966–76) (Spence, 1999).

The land revolution solved the Party's concern about mass mobilization as well as the necessity to extract unlimited human and physical resources for the war against the Nationalists. It can be seen that the reforms after the revolutionary victory were also the political tactics used to control the masses and establish a solid social and political basis of national unity. Through the reform, the Party realized its goal of overthrowing the old regime and reorganization of the grass-roots society, which laid the foundation for modernization. According to Mao, China's revolution took one form – through struggles to unite the peasants and create a united new nation (Gao, 2007). However, it is far too simple to judge the extent to which the reforms catered for the peasants' best interests, which may explain the partial failure of the People's Communes in 1960s, to be replaced by the Household Responsibility System (HRS) in the late 1970s. Esherick (1995) contends that the Chinese revolution was not liberation but the replacement of one form of hegemony with another. It had more to do with the alienation of Chinese society from an increasingly authoritarian state (Friedman et al., 2005, cited in Perry, 2008). Nonetheless, Mao's revolutionary path had its far-reaching implications for the Chinese government's current reform which faces the growing inequalities between the haves and have-nots. In this respect, lessons from the pre-revolutionary period have important implications for today's society (Perry, 2008).

Conclusion

The study of Chinese land reform history, characterized by a seeming cycle of land fragmentation, land concentration, land struggles and land re-fragmentation (redistribution) especially since the Ming and Qing dynasties, is instrumental to understanding the dilemma facing China's development today. It shows that the trajectory of land reform is the result of a long-term struggle between the state and the peasantry. Land has always been the driver for social and political changes. Political agendas underpinning the Ming, Qing and Nationalist regime reforms exemplify a major commonality – incremental change with a focus on economic resolutions to poverty – ended with certain failure to reorganize rural society for the benefit of the poor. In comparison, the road taken by the Communists was more radical and complete, but also ended with the failure to generate peasant incentives to develop the rural economy.

Social structures and organizations may explain the lynchpin of China's land reform and the constraints on peasant-centred land policy changes. Land has never become a catalyst for the creation of social space for poor peasants (Zhang, 2005). Rather, it is used by the state to exert stronger control over the sluggish economy and an increased threat from the local elite. This resulted in a loose social structure and organization that could not foster the collective force that would otherwise have been needed. The majority of the peasants continued to feel isolated from the mainstream economic and social organization while cultivating their tiny plots of land for survival. Furthermore, state-society relations are complicated by the interactions of various economic and political actors who pursue their own interests. This further complicates the way in which peasant interests can be safeguarded and relevant policy measures can be engaged.

This overview of land reform history reveals the indispensable exploration of economic and political reforms, especially those undertaken by Mao. The demise of the Ming and Qing dynasties has much to do with the characteristics of Chinese society underpinned by a realm of landlordism. That is why land continues to be controlled by the rural elites and state functionaries in dynasty after dynasty. Mao's attempt to dismantle this rural social fabric through radical land revolution and Peoples' Communes, however, embodied an institutional ineffectiveness in tackling its inherent nature of being traditional, bureaucratic, centralized, all-powerful and responsible to no one outside its ranks (Hinton, 1983). And lack of authentic representation for poor peasants in power and politics further explains the partial success of the revolution and land collectivization, as a state-dominated peasant movement is insufficient to safeguard peasants' rights and represent their interests (Wu, 2010). This also calls into question the extent to which the rural social structure was changed in the interests of those more disadvantaged groups.

This chapter demonstrates that land reform in Chinese history is inextricably linked with poverty and social inequality, which are embedded in the persistent dominance of state and local elites over the mass peasantry. This further pinpoints the need for political redressing of the fundamental issues concerning the lack of alternatives to rural development and agriculture in particular rather than a sole focus on land concentration issues. The latter were not the only factors for social and political changes. Other dimensions of poverty and inequality, that is, the non-landownership inequality factors such as high land rentals, interest rates, debts and natural resource constraints aggravated the tensions between different social groups and played a more important role in social and political struggles (Tang, 2006).

This has been inadequately addressed by all the regimes, although Mao's People's Communes seemed to mark a watershed from the past. Furthermore, individual peasants' vulnerabilities to domination by the elite would require concerted efforts to fight social inequality and poverty. This could be achieved through agricultural cooperation (Tawney, 1966). Yet, the drive to create efficient peasant organizations can be hindered by the power relationships between interest groups. China's land reforms have failed to create genuine institutions to counter the forces of local bureaucracy and political control. The state has managed to take the institutions into its own hands to re-establish new institutional orders very close to traditional ruling.

The demise of each dynasty and even the failure of the People's Communes underline the constraints to tackling the roots of poverty. When other causes of poverty were not well defined and tackled by Mao or by previous regimes, land became a relatively easy subject to be used as a medium for political gain. They all understood that they would not have gained effective political leverage had the land not been brought under their control. Mao's grass-roots-oriented strategy of 'putting politics in command' worked to serve his own politics, but did not stand the test of overcoming the persistent challenges of poverty and inequality – it was not just the land (see Burkett and Hart-Landsberg, 2005: 436). This argument might also explain why others interpret the revolution as irrational (Tsou, 2000), because better deals could probably have been struck between the revolutionaries and the rulers. This may also explain why the relatively 'soft' approaches of the Ming, the Qing and Nationalists did not accomplish what Mao did. In short, peasants have been agents of revolution in the sense of being operated as a machine to effectuate the preconceived political goals of the operators (Moore, 1967).

As land reform continues, it is important to enshrine the rights of the peasants whose organization and economic independence play a critical role in social and political changes. It is the challenge confronting Chinese society to create institutional alternatives that address the relations of production, society and development, which will benefit the disadvantaged groups. The lessons from Chinese history reveal that any change in political ideologies and actual political actions could become futile if the local constituencies are not given choice and power to engage in the reform process. It remains crucial to organize collective action to arrive at an accepted definition of the situation and a formulated programme for rural development (Fei, 1980).

References

Beattie, Hilary J. 1979. *Land and Lineage in China: A Study of T'ung-Ch'eng County, Anhwei, in the Ming and Ch'ing Dynasties*, Cambridge: Cambridge University Press.

Bianco, Lucien. 1986. "Peasant movements", *The Cambridge History of China*, 13, 270–329.

Burkett, Paul and Hart-Landsberg, Martin. 2005 "Thinking about China: Capitalism, socialism, and class struggle", *Critical Asian Studies*, 37 (3), 433–40.

Chao, Kang. 1986. *Man and Land in Chinese History: An Economic Analysis*, Stanford, CA: Stanford University Press.

Chiang, Tai-hsin. 1982. "A survey of early Qing reclamation policy and land distribution", *Lishi Yenchiu*, 5, 160–72.

Chinese History Textbook Net. 2009. *History Studies*, http://hist.cersp.com/kczy, accessed 23 January 2010.

Douw, Leo. 1991. *The Representation of China's Rural Backwardness 1932–1937: A tentative analysis of intellectual choice in China, based on the lives and the writings on rural society, of selected liberal, Marxist, and Nationalist intellectuals*, unpublished PhD dissertation, Leiden University.

Eastman, Lloyd E. 1988. *Family, Field and Ancestors: Constancy and Change in China's Social and Economic History 1550–1949*, New York and Oxford: Oxford University Press.

Elvin, Mark. 1973. *The Pattern of the Chinese Past*, Stanford, CA: Stanford University Press.

Esherick, Joseph. 1981. "Number games: A note on land distribution in pre-revolutionary China", *Modern China*, 7 (4), 387–411.

Esherick, Joseph. 1995. "Ten theses on the Chinese Revolution", *Modern China* 21 (1), 44–76.

Fei, Hsiao-Tung. 1980. *Peasant Life in China: A Field Study of Country Life in the Yangtze Valley*, London and Henley: Routledge & Kegan Paul.

Feuerwerker, Albert. 1983. "Economic trends 1912–1949", *The Cambridge History of China*, 12, 28–128.

Friedman, Edward, Pickowicz, Paul and Selden, Mark. 2005. *Revolution, Resistance, and Reform in Village China*, New Haven: Yale University Press.

Gao, Wangling. 2007. "土地改革: 改天换地的社会变动" (Land reform: A complete social movement), *China Rural Studies*, http://www.ccrs.org.cn, accessed 15 December 2009.

Heijdra, Martin. 1998. "The socio-economic development of rural China during the Ming", in Mote, F.W. and Twitchett, Denis (eds), *The Ming Dynasty, 1368–1644, Part 1, The Cambridge History of China*, Cambridge: Cambridge University Press, Vol. 8, 417–758.

Hinton, William. 1983. *Shenfan*, New York: Random House.

Ho, Ping-Ti. 1962. *The Ladder of Success in Imperial China: Aspects of Social Mobility, 1368–1911*, New York and London: Columbia University Press.

Hsiao, Kung-Chuan. 1967. *Rural China: Imperial Control in the Nineteenth Century*, Seattle and London: University of Washington Press.

Huang, Philip C.C. 1985. *The Peasant Economy and Social Change in North China*, Stanford, CA: Stanford University Press.

Lardy, Nicholas R. 1983. *Agriculture in China's Modern Economic Development*, Cambridge: Cambridge University Press.

Li, Wenzhi. 2007. *The Loosening of the Feudal Land Relationships in the Ming and Qing Dynasties*, Beijing: Social Sciences Academic Press (China).

Lin, Jingrong and Chen, Zhenzhuo. 1995. "浅谈清代的土地制度" (A brief introduction to the land system in Qing Dynasty), *Fujian Forum*, 3, 14–17.

Menzies, Nicholas K. 1994. *Forest and Land Management in Imperial China*, New York: St. Martin's Press.

Michael, Franz. 1966. *The Taiping Rebellion: History and Documents, Vol. 1: History*, Seattle, WA: University of Washington Press.

Moore, Barrington. 1967. *Social Origins of Dictatorship and Democracy: Land and Peasant in the Making of the Modern World*, Boston: Beacon Press.

Perkins, Dwight. 1969. *Agricultural Development in China, 1368–1968*, Chicago: Aldine.

Perry, Elizabeth J. 2008. "Reclaiming the Chinese Revolution", *Journal of Asian Studies*, 67 (4), 1147–64.

Riskin, Carl. 1987. *China's Political Economy: The Quest for Development Since 1949*, Oxford: Oxford University Press.

Schiffrin, Harold. 1957. "Sun Yat-sen's early land policy: The origin and meaning of equalization of land rights", *Journal of Asian Studies*, 16 (4), 549–64.

Spence, Jonathan. 1999. *The Search for Modern China* (2nd edition), New York: Norton.

Tang, Zongli. 2006. "Land distribution in Mao's investigations: Poverty and class struggle", *Journal of Contemporary China*, 15 (48), 551–73.

Tawney, Richard Henry (ed.). 1939. *Introduction to Agrarian China: Selected Materials from Chinese Authors*, London: Allen & Unwin.

Tawney, Richard Henry. 1966. *Land and Labour in China*, New York: M.E. Sharpe.

Tsou, Tang. 2000. "Interpreting the revolution in China: Macro-history and micro-mechanisms", *Modern China*, 26 (2), 205–38.

Wang, Yeh-chien. 1973. *Land Taxation in Imperial China, 1750–1911*, Cambridge: Harvard University Press.

Wilkinson, Endymion. 1978. "Introduction", in Jing Su and Luo Lun (eds), *Landlord and Labour in Late Imperial China: Case Studies from Shandong*, Cambridge and London: Harvard University Press.

Wolf, Eric R. 1973. *Peasant Wars of the Twentieth Century*, New York: Harper Torchbooks.

Wong, John. 1973. *Land Reform in the People's Republic of China: Institutional Transformation in Agriculture*, New York: Praeger Publishers.

Wu, Jinglian and Reynolds, Bruce. 1988. "Choosing a strategy for China's economic reform", *American Economic Review*, 78 (2), 461–66.

Wu, Li. 2010. 农民意识，家庭经营，土地制度：读毛泽东视野中的中国农民问题 (*Peasants' Awareness, household economy and land system: Review of the problems of the Chinese peasantry in Mao Zedong's views*), http://www.21ccom.net/articles/zgyj/ggzhc/article_2010103023272.html, accessed 15 December 2010.

Wu, Shang-Ying. 1955. *Sun Yat-Sen and Land Reform in China*, reprinted from the Henry George News, March 1995, http://www.cooperativeindividualism.org, accessed 21 November 2009.

Zhang, Hongming. 2005. *Land as Symbol: The Restudy of Lu Village*, Beijing: Social Sciences Academic Press (China).

Chapter 4

China's Land Tenure Reform
since the 1970s: Crossing the River
by Touching the Stones?

Introduction

Land reforms after the founding of the People's Republic in 1949 took three major stages. The first stage saw the emphasis on 'land to the tiller' as an historical watershed from the past two thousand years of failed reforms in bringing equality of landownership to the fore. Consequently, increased incentives to peasants in grain production contributed significantly to agricultural development. However, concurrent to this positive outcome was the emergent inequality between the poor and rich peasant groups as well as the challenge to provide the needed public services in the countryside. To tackle this problem, the second-stage land reform launched in the 1950s was aimed to avert land amalgamation derived from enlarged inequality in order to ensure the sustainability of the 'land to the tiller' principle and practice. As a result, the system of People's Communes was institutionalized to enshrine peasant collectives in landownership and management. However, land management was not fully coupled with sound democratic practices to ensure ample participation of the peasant members, which led to low productivity in grain production but improved public service delivery (Wang, 2009).

By contrast, the third stage of the land reform, characterized by the introduction of the Household Responsibility System (HRS) in the late 1970s, was a tactical move towards more incentive-based land management structures by combining the features of collective landownership and long-term rights of individual households in land use and management with a view to offering peasants greater incentives in farming. Hence, it is an institutional innovation in balancing the interests of the state, village collectives and households in drastically enhancing peasants' interests and rights in agriculture, thereby ensuring food security. China saw its agricultural growth rate of 4.4 per cent from 1979–2008, which contributed to its success in feeding 20 per cent of the world's population with only 8 per cent of the world's arable land (Wang, 2011: 35). Paradoxically, the HRS has not contributed to scaled farming and production and has contributed to inadequacy of public service provision (Wang, 2009). The share of agriculture in GDP declined from 28 per cent in 1978 to 10.6 per cent in 2009, with employment share declining from 71 per cent to 39.6 per cent. As a result, China has become a net importer of foodstuffs since 2004 (Wang, 2011: 37).

Furthermore, the HRS has displayed its weakness in sustaining land use as evident in rising farmland loss-induced social conflicts due to the lack of effective collective action by the peasants to use the land sustainably and to defend their rights (He, 2010; Zhao, 2010). Recent land law and policy developments in the use of neo-liberal or pro-market orthodoxy underlines the assumption on the linkages between strengthening peasants' land rights and their entry into the emerging land market. As such, it is argued that the system of land transfers should be drastically improved to enable those peasants with no interest in land to sub-lease their land use rights to those who would otherwise invest in the land. This is deemed necessary to facilitate land use efficiency and scaled production, which would embody the consolidation and continuation of the HRS. In other words, an incremental approach is needed to allow peasants to leave their villages for the cities in order to facilitate a more efficient and scaled rural economy (Wang, 2009). The latter is often deemed essential to combating poor land governance and contributing to equitable development. It is taken for granted that peasant smallholders would attain a stronger bargaining position vis-à-vis the local state over issues of farmland acquisition, expropriation and allocation of benefits once they have been given more discretionary power. However, policy measures underpinned by these theories have yet to produce desirable outcomes.

Can the idealistic theory characterized by free and fair land transfer mechanisms (not yet put into policy), under the banner of rural collective landownership to enshrine peasants rights pave way for rapid agricultural development and urbanization? Contrastingly put, would not it lead to land concentration in the hands of the mighty few? Would land smallholding be inherently improper for scaled production or would it be more sensible to develop it further for the best interests of the peasant households? And would urbanization be the unstoppable trend against the Chinese reality of overpopulated cities and endangered natural resources? Lastly, would there be no alternatives but to driving peasants off their land for the sake of a so-called scaled economy and urbanization? Whose scaled economy and urbanization?

China's land tenure reform agenda bears resemblances with other countries that prioritize land tenure security as a panacea to all the social and political ills obstructing fast economic development, and not only in the countryside. However, land tenure security does not have direct causal links to economic development (Ubink et al. 2009). A lack of understanding of the social and political dimensions of land tenure among policy designers is often claimed to be responsible for policy and institutional failures. Thus, a more flexible approach to designing land tenure systems and policies concerning land use planning in line with local concepts and practices is advocated as a third way alternative to the one-size-fits-all models in the international context (Otto, 2004, cited in Ubink et al. 2009).

Despite the ongoing debates and policy experimentation, the myth of land tenure has not been creatively addressed. It is obvious that one cannot strengthen land tenure security through pure administrative and legal means. It is more important to understand its underlying structural determinants rather than focusing on its narrow domain, as land tenure is an integral part of development and governance processes (Zhao, 2010). Thus, the design of land tenure systems from a pro-poor angle needs to be brought into the current debates on land policy trajectories in China. As China has reached a critical stage of development characterized by increasing social inequality, chronic poverty and natural resources depletion, this paradigm shift is more needed to achieve sustainable rural development outcomes than the one-sided preference over land and property rights.

This chapter first illustrates the case of land tenure-induced constraints to sustainable development in China. This is followed by a brief overview of China's land policy responses since the late 1970s, mapping out the trends and analysing the underlying issues of poor land governance. It then discusses the major international and national debates on land tenure reform to illuminate the Chinese case. It concludes with an emphasis on the need to rethink the current state-led pro-market land policy approaches and for the ways forward for land reform in China in learning lessons from other countries relevant to the Chinese context.

Challenges of Sustainable Natural Resource Management

China is facing daunting challenges of sustainable natural resource (land, water, forest, grassland) management, which set constraints to its pursuit of development goals. There has been rapid urbanization, for instance, that has expanded over 70 per cent of the Beijing-Tianjin-Hebei region during 1990–2000. Despite its inducement of economic growth and the provision of better opportunities for the peasants, rampant urbanization has also caused huge losses of farmland. By 2000, China's total arable land area was reduced to 128 million ha, equivalent to 0.11 ha per capita – less than half of the world's average of 0.23 ha. In the last decade, it was estimated that 1.5 million peasants lost their land, and in 2005, the number of landless or unemployed peasants had increased by 3.8 million (Tan et al., 2005: 187–8; Liao, 2007: 163). According to the Ministry of Labour and Social Security, by 2006 the total number of landless people had reached 40 million with an annual rate of increase of 2.65 million due to land acquisition (China Governance Information Centre, 2006: 30). Unjust land acquisition processes pose a threat to sustainable livelihoods – a major cause of social instability.

Land desertification is expanding at an alarming rate; by 1999 it had encroached on 18.2 per cent of the entire mainland, and this trend coupled with vegetation loss has not been effectively reversed to date. China is also one of the countries with severe problems of water shortage and pollution. The available water usage per capital in its 14 provinces and municipalities has gone far below the international minimum line, and about two-thirds of the Chinese cities face water shortages (Zheng, 2004: 32). The forest coverage by 2008 was 18.2 per cent, far below the world average of 29.6, although gigantic afforestation programmes have offset forest degradation and depletion to a limited extent. By the end of 2007, one-third of China's rangeland had been degraded with severe impact on the livelihoods of peasants and pastoralists (Bao, 2009: 137). In addition, to a certain extent, improper uses of farmland, such as the overuse of fertilizers to increase grain production, has caused the pollution of the land and waterways – a direct threat to food safety (Huang et al., 2006).

These natural resource constraints reveal an inconvenient truth. The northern provinces with a less developed economy than the south have the largest farming areas in the face of unfavourable climatic and natural conditions such as water shortages, soil and wind erosion. The southern provinces where climatic and natural conditions are more suitable for farming, by contrast, have experienced a more severe level of farmland loss and decline of agricultural productivity. In addition, agriculture has no longer been a priority for most regions along the eastern coastlines, which received the preferential policy treatment to 'get rich first' dating back to the start of the market reform. As a result, peasants were encouraged to migrate to the cities for the economic boom, or they joined local enterprises that did not always have to do with land (Chen, 2009).

To a large extent, the majority of the Chinese peasants depend on farming-related activities in the face of ecological vulnerability and external threat with regard to land evictions, all of which have contributed to the persistence of poverty, especially at a time of world economic crisis. In 2008, annual net rural income per capita for peasants was US$697, which shows that China remains a developing country, and it will become more difficult for the peasants to attain further income increases in the coming years.[1] In particular, it would be demanding to maintain the current level of grain production. Any unexpected severe flooding or drought may cause heavy losses of natural assets, which threatens the fragile rural economy (Sheng and Bai, 2009). All these challenges are compounded by the enlarging gap between the rich and poor and between urban and rural localities.[2]

China's rapid economic growth has also been a major factor in the loss of farmland and the degradation and depletion of natural resources. Land acquisition has contributed to a major source of local extra-budgetary revenues. From 1999 to 2009, the annual increase of revenues from farmland acquisitions reached 43.2 per cent, and thus constituted 43.7 per cent of total local government fiscal revenues in China. Even for the less-developed western hinterland, studies have found an alarmingly increased proportion of local government fiscal revenues related to land (Man, 2010: 16–18). These lucrative deals often link to the local governments' manipulation of farmland transfer procedures, where farmland is acquired at low prices and given to businesses at a much higher value without just compensation to the peasants. These practices pose a severe threat to the landed environmental and resource sustainability (Han, 2010). No wonder better land management systems

---

1   In the World Bank's World Development Report 2008, China is listed as a lower middle income country. In 2004, 9.9 per cent of the population lived on less than $1 a day, and 34.9 per cent on less than $2 a day (World Bank, 2007: 336).

2   For instance, the urban-rural income ratio of 3.22:1 in 2005 is an indicator of the increasing social inequality, see (Zhu and Prosterman, et al., 2006: 764).

and improved rural governance, based on grass-root democracy and peasants' self-governance of village affairs, are needed to offer some solutions to these problems (Wang, 2009).

An Incremental Approach to Land Policy Changes

The post-Mao era saw a reformist vision for China's development. Seemingly endorsing neo-liberal approaches, the policy for land de-collectivization characterized by the HRS seemed to work in the early stage of the reform, which resulted in increases in agricultural production and poverty reduction (Oi, 1999). Despite its initial success, the HRS has not enabled agriculture to substantially lift the majority of peasants out of poverty, although other factors also count. The major land policy changes are outlined in Table 4.1:

Table 4.1   China's land policy changes in the post-Mao reform era

1978–1986: The Household Responsibility System (HRS) replaced collective farming in several regions. Rural households have land use rights under collective landownership.

1998: Land Management Law to uphold the limitation to rural land subcontracting and transfers.

2002: Rural Land Contracting Law to protect the contractual use rights of households and open the door to farmland market, allowing for land use rights leases, exchanges and transfers under collective ownership without changing original farmland usage. Involvement of non-villagers was strictly limited.

2007: Property Law as the first law to explicitly offer protection for private property rights. Farmland remains the property of the village collective.

2008: Central Party Committee (CPC) Decision on Major Issues Concerning the Advancement of Rural Reform and Development: further call for farmland transfer, lease, exchange and swap based on market mechanisms and peasant consent and in order to enhance scaled farming and peasant incomes under village collectives. Pilots on trading of collectively-owned construction land without first going through government acquisition were given the green light.

2010: CPC Opinions on Scaling Up Integrated Urban-Rural Development (No. 1 Document): Accelerating contracted land use rights transfers for scale economies; strengthening land management through registration of contracted land use rights; planning to complete registration of collective-owned land within 3 years.

2011–2012: Land Management Law under review, new version to be promulgated.

Source: Author's own compilation, based on relevant laws and policies and Tong and Chen (2008).

Table 4.1 shows that the government has taken an incremental approach to changes in land laws and policies in a more market-oriented fashion. Furthermore, the strengthening of land law enforcement characterizes a major agenda on farmland protection and local governance accountability. The government claims that it has put in place the world's most rigid laws to contain farmland loss, although it is hard to justify this. The 1998 Land Management Law, for instance, stipulates that only the State Council and provincial government have the right to approve land acquisition plans. It also grants peasant households 30-year land use rights backed by extendable written contracts. However, as briefly indicated, these measures have been largely ineffective due to local government manipulation, among other reasons, which is self-evident in the rising cases of social unrest related to the infringement of peasants' land rights. This may explain the aim of the 2002 Rural Land Contracting Law and 2007 Property Law, which provide a foundation for fully-fledged land markets as seen in the call for land registration to facilitate peasant discretion on land use rights transfers or sub-leasing – not land sales.[3]

For land evictees, these laws and policies have yet to put in place clear and detailed articles on transparent and accountable land acquisition processes to allow them to have ample access to government information. Neither have they detailed stipulations on the conditions under which forced removal is allowed. This further disadvantages the peasants whose legal awareness and power to exercise their legal rights remain weak. As a result, land acquisition in the name of economic development has become unstoppable. And local governments lack the political will to correct their own mismanagement of land resources. The latest land-related policies seem to grant the village collective more leeway in transferring their non-arable construction land, which might trigger more discontent of local government with the central government, as the former's power can be undermined (Tong and Chen, 2008).

It seems that the government only has one last resort to tackle the above issues, that is, putting in place more and more policy implementation measures to further strengthen peasant land rights and to create more incentives for them to enhance land investments through the introduction of the land market. The promulgation of the 2007 Property Law, for instance, has been claimed to be a landmark victory for the advocates of private property. To some extent, the Central Party Committee's October 2008 rural reform policy is a step forward to allow peasants to lease or transfer their land in order to raise rural incomes and facilitate rural–urban migration under the assumption that it would enable them to have substantial decision-making power over their land. This move seems to resonate with those who follow the de Soto approach to capitalizing on land and its associated rights

---

3   Farmland can only be transferred from the hands of peasant households to the local government and developers if it is turned into construction land (for instance for infrastructure building) first by the local government, in addition to meeting relevant land use plans and government approval procedures.

for rural economic development (de Soto, 2000), as they believe that it would lead to the formation of larger and more efficient farms and thus the elimination of inefficient family farms to stimulate the rural economy.

To follow suit, for instance, the promulgation of the Land Registration Methods by the Ministry of Land and Resources (MLR) in 2008 is meant to streamline land use management between different line agencies and to help solve land disputes over jurisdiction boundaries among villages. The registration of collective ownership of farmland and construction land has been a primary task for the MLR, while the Ministry of Agriculture (MOA) is responsible for the registration of farmland contracting rights. Registration certificates are issued to the collective on its title as an overarching responsible entity that is encouraged to invest in land shareholding arrangements. This policy also puts forward articles on land rights protection allowing for a transparent land registration process in which registered materials can be openly accessed. It was reinforced by the 2010 No. 1 Document that stresses the task of completion of collective land registration within three years in order to improve land management. But the implementation of this policy may not contribute to the needs of individual households. As a leading land policy expert contends, 'this only serves as an administrative tool in terms of clarifying village boundaries rather than reflecting the changing needs of peasants for land uses due to the lack of resources and capacity of the government in developing more socially-oriented land use monitoring mechanisms'.[4]

However, these strengthening measures have not always worked, and it is normal for them to have negative effects on sustainable land use and social equity. For instance, a recent policy calling for striking a balance between farmland acquisition and reclamation (*zengjian guagou zhibiao zhihuan*) is a case in point. In essence, this policy requires local governments to make farmland available through land reclamation prior to the acquisition of existing farmland in order to maintain the overall balance of farmland acreage. In practice, this has provided the local governments with the opportunity for land acquisition on a more substantial scale. Rapid local experimentation is under way through the demolition of peasants' houses, the land of which is then reclaimed for new farmland without knowing how suitable the land is for farming. By doing so, the existing farmland can be acquired for the development of scale economies – mostly for large-scale agricultural or even non-agricultural-related investments by big households or corporations. As a result, many peasants are losing their housing land and asked to move into buildings close to the cities in the name of urban-rural integration (*chengxiang yitihua*). However, their livelihoods are adversely affected due to a lack of appropriate arrangements for their compensation and social welfare. Even if just compensation is made to them, many of them are unwilling to give their land away as they may live in poorer conditions in the cities where the equalization of public service provision remains a huge challenge. Local governments, however,

---

4   Informal interview with a senior land expert from the China Land Survey and Planning Institute in June 2008.

see this move as a necessary and probably a sole step towards more rapid economic development to increase their local revenues.[5]

This new round of *de facto* land enclosure practices by local governments obviously shows the ineffectiveness or failure of relevant land policies, laws and institutions from the central government. It is also related to the complexity of local governance, for instance, the 'mystic' role of the village collective in the land acquisition or even illegal eviction process. While being weak in safeguarding their members' rights and interests, the village collective connives with the local state and businesses in striving for lucrative land deals without the consent of their constituencies in many cases. Local courts serve the needs of the government rather than the interests of the evictees as sometimes they are directly involved in illegal evictions. Where access to effective legal resources is lacking, the peasants have to take the law into their own hands, as seen in cases where they burned themselves down or confronted the government by force.

There is a lack of effective monitoring systems put in place for local practices, and thus it is far from being clear how local governments carry out farmland acquisition procedures without ample space for peasant participation. The current development trend towards urban–rural integration is a prominent development agenda for the accomplishment of a harmonious society, and is being experimented with use of market-oriented mechanisms for the transfer of farmland contractual rights. Yet, the role of the local state in the rural land market that does not actually exist is so complex that poor peasant households are always disadvantaged. Hence, relevant polices related to this agenda have not proven effective or pro-poor at all, and in some cases, have exacerbated farmland loss and land-induced social conflicts (see Zhou, 2007; He, 2010).

It can be seen that the livelihoods of the Chinese peasantry continue to be at the mercy of economic development in general and land tenure insecurity, farmland loss and chronic poverty in particular. The limitations of all those administrative and legal measures in curbing corruption in land governance and safeguarding the interests of the poor households are far less effective than expected (Van Rooij, 2007). Local governments may not even have the intention to follow the central government's mandates, but are paying lip service to policy implementation. As a result, land policies aimed at strengthening the rights of the poor and undermining the power of the local government and businesses have backfired. In essence, what impacts these policy developments are exerting on the poor, how the latter react to the developments, and what other reform measures are needed to make land sustainable for the poor, need to be further explored for an improved

---

5   Various local experiments on this new round of land acquisitions and evictions is discussed in 'The Chinese villages under siege: Demolishing peasants' houses and misused policies' (English translation) on South China News Net (nanfang xinwenwang), 15 October 2010, see http://www.southcn.com. For more information on the rural-urban integration policy, see http://www.chinanews.com.cn/cj/cj-plgd/news/2010/01-31/2101018.shtml

understanding of the complexities of land tenure as interwoven with development and governance reforms.

A Knowledge Gap between Theory and Practice in Land Tenure Reform

To further understand China's land tenure reform, as illustrated above, one should not only focus on land and property rights. How land is governed by the local government and how it is inextricably linked to peasants' livelihoods and sustainable land use deserves in-depth analysis. Current schools of thought have not paid enough attention to the importance of understanding these linkages. Neither is there sufficient literature written by Chinese scholars on these issues. Although it is beyond the scope of this chapter to detail these issues, it attempts to map out the main arguments primarily illustrative of current Chinese scholarship which has an increasingly prominent role to play in Chinese politics. Linking with lessons from land reform in other countries, this approach thus is aimed at filling the current theoretical gap, but more importantly, provides a framework on which the ensuing empirical chapters are based.

Market-oriented Land Tenure Reform

To a certain extent, international land tenure reform is driven by the need to promote pro-market approaches that can facilitate fast economic development. Land tenure is often referred to as 'the perceived right by the possessor of a land parcel to manage and use the parcel, dispose of its produce, and engage in transactions, including temporary or permanent transfers, without hindrance or interference from any person or corporate entity, on a continuous basis' (Migot-Adhola and Bruce, 1994: 3). Putting it into practice, land titling and registration programmes have been widely implemented with donor support in many developing countries with both positive and (probably more) adverse impacts on the livelihoods and rights of the poor. As Lund contends, when private property is taken as the highest level of tenure security, unintended consequences of these programmes may occur. Tenure security for one group may correlate with decreasing security for the other (Lund, 2000: 16).

In a similar vein, key discussions on China's land issues are focused on the lack of clarity and transferability of rural landownership as a hindrance to the realization of scale economies in rural China. The current hybrid land tenure structure is widely recognized as ambiguous and weak in nature and unfavourable to equitable land transfers. As a result, local elite capture and corruption in land expropriation are tantamount to poor land governance (Ho, 2005; Wang, 2009). Moreover, in the wake of the introduction of the HRS, rural collective power with regard to sound village administration and land management has weakened. Thus, Wang (2005: 73) points out that the most practical path for future land law reform should answer who the real landowner is and how the owner can exercise the rights. A failure to do so has contributed to locking up the peasants on their tiny parcels of land, which is economically unproductive for meeting the possible long-term needs of the country. A solution lies in improving the land rights structure through establishing a transparent and efficient land market in rural areas (see also Schwarzwalder, 2001). In addition, political reform should be coupled, as Wang (2005) further argues, for the deepening of rural democratization to allow peasants to choose their own leaders in a better way so that decisions on their land can be made by those who represent the peasants' own interests.

These contentions are followed by the more liberal experts calling for land privatization. For them, land reform in China is at a crossroads. Swift public action is needed to allow peasants to extend their land rights to buy and sell land freely, which will contribute to combat illegal land seizures and to build an orderly land market under the rule of law. This approach is also expected to address the concerns of some economists over a lack of impetus for rapid agricultural development, for they believe that the current land tenure system sets limits to technological advances in agriculture by obstructing land concentration or the maximization of land use efficiency. To improve agricultural productivity, it is necessary to entrust the peasants with the rights to sell, subcontract or merge their land with others in shareholding companies. For instance, Zhu and Prosterman et al. (2006: 834) assert, 'China should consider going beyond a tenure system of thirty-year rights by either providing farmers with full private ownership rights to land, or nationalizing agricultural land and giving farmers perpetual use rights'. In a similar vein, Pieke (2005: 107) affirms that the one-size-fits-all land policy is not suitable for China, especially for those relatively developed regions where agriculture plays a minor role in rural development. As many peasants are not allowed to dispose of their land, they continue to keep their land under grain rather than other uses. This economic inefficiency in land use contributes to the developmental stalemate currently confronting regional agriculture. Pieke suggests that the peasants should be allowed to freely mortgage or sell their land use rights to raise money for commercial ventures or other purposes (see Benjamin and Brandt 2002; Deininger and Jin, 2005). These views resonate largely with de Soto's stance on private property ownership as essential to economic development (de Soto, 2000).

Although strengthening individual peasants' land and property rights holds the key to sound land governance in China, it does not justify the need for either land privatization or nationalization. Or simply put, it may again become a one-size-fits-all approach which certainly does not suit many diverse development contexts. In fact, many Chinese peasants oppose privatization or even extended land tenure structure as long as they enjoy overall income security under collective landownership. Moreover, the heterogeneous ways of land management they deploy do not invite the intervention of sweeping policies (Qiao, 1997; Rozell et al., 2005). Were land to be re-nationalized, the market economy that China has adopted would be pointless. Were it to be privatized, one should not overestimate its potential advantages. According to Wang and Xu (1996: 202–3), land privatization would incur inherent problems. First, it would not drive agricultural modernization, since it would tie individual peasants to their fragmented land that hinders large-scale farming, at which the mainstream policy is aimed. Second, it could lead to changes in the existing rural land relations with the likelihood of re-emergence of landlordism. This would lead to poverty and deprivation of the majority of peasants. Only under public or collective landownership can the majority of peasants be protected from exploitation, and can social equity be realized to ensure common prosperity. In order to improve land productivity and agricultural growth, there is a need to realize economies of scale in agricultural production through collective means to ensure social equity for all the peasants. Bromley (2008) agrees that exclusive individual landownership is not essential to agricultural development. Instead, village organizations and property relations should ensure economic benefits for the poor. Collective landownership provides the conditions that enable peasants to move back and forth to their land in response to changing conditions in the wider economy.[6] According to Chen (2009), China may need to simulate the experiences of Japan, South Korea and Taiwan by organizing specialized cooperative organizations on the basis of the HRS. By organizing peasants in agricultural production and marketing, the cooperative is seen as a more effective institution to achieve the goals of scale economies than the institutionalization of private landownership.

Additional factors add complexity to the debates on land tenure security. It is a fact that the majority of Chinese smallholders still confront poverty and social inequality brought about by imbalanced economic development. When other economic opportunities arise, some peasants may forgo their land to pursue the alternatives. With little land in their hands, leaving the land to the local government and businesses in return for compensation is not always a bad deal, especially for young people who are more likely to invest in local businesses or simply to migrate to the cities. With few alternative opportunities offered except for farming in the countryside, many peasants, especially those in relatively developed peri-urban areas, desperately need a quick relief from poverty no matter what opportunities are offered. All these factors are actually conducive to land acquisition by the local state. While this may support the neo-liberal lines of thinking, the mystery remains as to whether rural–urban migration is indeed conducive to the peasants or serves the sustainable development needs of both rural and urban areas, especially as it is simply not possible for the Chinese cities to accommodate more and more migrants. At this stage, few facts or scenario analyses can be found to convincingly support the thesis of rural–urban migration. This is vividly exemplary of the incremental approach to Chinese policy-making and implementation or the nature of the whole economic reform characteristic of trial and error or 'crossing the river by touching the stones'.

In a nutshell, there are no causal links among private landownership, scale economies and sound land governance. As the market mechanisms are taken as

---

6   This is further illustrated in Chapter 7.

prerequisites to rapid rural economic and agricultural development in China, the Chinese trajectory of land management does not display major differences from many other developing countries. Where China deviates from them lies in China's constant experimentations with strengthening market factors in the hybrid land tenure structure to boost the local economy, which could lead to the formalization of the land market in the foreseeable future.[7] However, the role of the state in market–peasant relations is so complex that any idealistic support for strengthening of property rights could be futile. It is in this context that China's reform has not addressed its own social and political challenges that undercut its success.

Socio-political Embedding of Land Reform

As is already known, in post-socialist countries (mainly in the Eurasian region), the introduction of private property rights has brought about the breakdown of the earlier cohesion of village life with its often elaborate, though informal, structure of rights and obligations (Myrdal, 1968; Todaro, 2000). The Chinese case of pro-market land reform is part of this process. From a socio-political point of view, the adverse effects of the reform have received scant examination in research and policy debates.[8] The HRS being transformed along market lines underpins the changing social and political relations as critical challenges for sustainable livelihoods and pro-poor governance in the vast countryside. As Hann et al. (2003: 23) indicates, the HRS has caused inefficiencies and environmental degradation to the 'social–technical system' of pastoral societies in certain regions. Hann goes on to argue that it exemplifies simple models of ownership that favour either individualism or communalism. Paradoxically, even in modern capitalist states, no absolute property rights or ownership systems are found. 'Rights to use and to manage are often exercised by parties other than the legal owners. The power to control, or simply the power of access, is of greater practical significance than legal ownership' (Hann et al., 2003: 24). Aside from the power issue, it is important to understand the multifunctional relationships vested with property and the underlying ideology, culture, legal regulation and property practice. This viewpoint can help us to understand the negative consequences of land privatization in those countries where emergent social conflicts and perverse economic impacts have become paramount. A sole emphasis on the introduction of a land market and efficient land administrative institutions in these countries thus proved to be unrealistic (Hann et al., 2003: 25–7).

As such, decollectivization may not bring about the expected changes in property relations, since the prescribed land policies and laws depend not only

---

7   Informal interview with a senior land expert at the Chinese Academy of Social Sciences in April 2009. He pointed out that China's land reform was directed towards western-style land privatization, although China had a long way to go and the government would not like to call it given the political sensitivity of the land reform at this stage.

8   This is reasonable given the difficulty in conducting relevant fieldwork.

on economic factors, but also on local perceptions of morality and the informal rules of the community (Finke, 2000). This is evident in many cases where local communities are not purely driven by the material conditions of the mode of production, but by the need to forge or maintain stable forms of relatedness in an unstable environment. Reliance on traditional collective institutions to realize this need plays an important part of their way of life (see Brandtstadter, 2003). Moreover, decollectivization has brought about negative effectives on rural living conditions and agricultural efficiencies, which are also becoming more apparent in the Chinese case. A cautionary approach needs to be developed in order not to impose a pre-conceived reform agenda on the local community whose views and capacity of self-organization have to be well understood first (Hann et al., 2003).

The Chinese case reveals the embedding social and political complexities in the HRS and the wider development challenges. In remote poor areas in particular, land is primarily used for subsistence by the majority of peasants, which requires its equal distribution to accommodate demographic changes. Land readjustment is not a common practice to this end. Many peasants in these areas may not show strong interest in receiving land use contracts to secure land tenure security. They are more concerned about how to make the land meet their basic needs rather than seeking economic and political rights embedded in their land. To a certain extent, land is not always seen as a lucrative asset, as peasants did not want to bear land-reduced taxes and fees imposed on them especially before 2006 (Zhao, 2008b). The village administrative allocation of land has been strongly criticized for its negative impact on peasant incentives in land investment, and land tenure security through the registration of peasant land rights is strongly advocated (Zhu and Prosterman, 2006). The current legal framework sets restrictions on the practice of informal land adjustments among the village households, but it has been ineffective and cannot tackle the social complexities inherent in land relations (Zhao, 2008b).

For indigenous communities living on the margins of development in many remote and mountainous regions, property rights carry an exotic meaning. Some groups value their communities as the safeguards of collective resources, communal land projects and equitable distribution of resources. Property undercuts a relationship between people, embedded in a cultural and moral framework, and their own vision of community (Hann, 1998). For instance, for pastoralists in Inner Mongolia, rangeland is managed in line with communal rules developed over the course of their history, which does not favour the delimitation of the land each household uses as mandated in the HRS. It is found that the latter does not mitigate but actually contributes to the 'tragedy of the commons'. Thus, new policies to strengthen tenure security and to promote agricultural production may conflict with the systems of the vulnerable communities as evident in the fragmentation of community relations and land degradation. In this context, local communities need to be given the rights to utilize their resources in ways that best suit their own interests through the establishment of small-scale collective property systems (Sturgeon, 2004; Li et al., 2007; Yang, 2007).

Chinese land policy developments send a confusing or conflicting signal. On the one hand, strengthening the HRS is upheld as seen from an increased emphasis on individual land and property rights. This seems to undermine the power of the village collective. On the other hand, collective institutions such as the formation of land cooperatives are experimented with to stimulate scaled production and thus overcome inefficiencies of small-scale landholding. Although land cooperatives are meant to be based on members' free will and initiatives, as Chapter 6 shows, the outcomes have turned out to be the opposite. Thus, the dilemma facing policymakers is how to find the proper match between individual and collective land management given their advantages and disadvantages, while maintaining the status quo of a socialist market economy or 'socialism with Chinese characters'. Whereas village collectives and local governments continue to play a dominant role, how to enable the participation of individual peasant households in land management remains unsolved. In pursing scale rural economies, this situation can even be exacerbated by land consolidation programmes that further fragment social and political cohesion and trigger enlarging land inequality and landlessness (see World Bank, 2007).

Furthermore, the conventional pro-market approach characterized by the formalization of land rights by the post-socialist countries and China (although China has yet to implement land privatization) adds complexity to village governance. According to He (2010), few peasants in his study sites that spanned over 20 provinces and municipalities expressed their inclination for private landownership. The peasants' need is more about how to revitalize the collective for the common good of the peasants through organized farming-related activities. Moreover, the HRS has not contributed to the empowerment of the poor, as the marginalization of the peasantry by the local elite continues to deter peasant incentives in land investment. And lack of social capital and access to various economic and political resources contributes to their lack of institutional choice to enhance their low economic, social and political profiles. In the face of absent or ineffective collective action by the peasants in their daily struggles for poverty alleviation and improved village governance, peasant support for the implementation of government-led development programmes has shown a tendency for disinterest (Zhao, 2008b). Consequently, any attempts to privatize the land or to simply rely on the market to overcome the underlying social, political and economic obstacles can be misleading. As along as the social and political rights of the poor are not enshrined in law and enforceable, it is unrealistic to pursue the course of land privatization (see Van Rooij, 2007).

Pro-poor Land Tenure

'Pro-poor land tenure' seems to be a fashionable term invented by non-Chinese, but in principle Chinese land and overall development policies are ostensibly pro-poor, as seen in their incessant emphasis on equitable development and social management for building a harmonious society. However, in fact, the challenges for successfully implementing these policies have to confront the enlarging inequality between the rich and poor groups and between rural and urban areas in hugely disparaged development outcomes.

The current dualistic institutional setup for rural and urban development is based on unequal allocation of resources that favour the cities, which is thought to be the cause of rural backwardness and weakened land and property rights of the peasantry. In essence, it is argued that as long as rural migrants in the cities cannot enjoy the same social and economic rights as the city dwellers, the speed of land transfers and urbanization will be severely undermined. Moreover, as fiscal arrangements are insufficiently tailored for the rural areas, rural public service delivery is put at risk, and peasants are prone to land expropriations in the wake of strong incentives for local governments in pursuing their economic targets. This dualistic institutional structure, when applied to farmland acquisition, often leads to lower compensation standards for the peasants than for city residents, who can sell their houses freely on the market. Thus, this unequal system ought to be abandoned in an incremental way with the deepening of the market-oriented reform (Wang, 2009).

It is important to note that gaining more economic, social and political rights for the Chinese peasantry is not the only parameter for societal transformation and the goal of development with equity (see Li and Bai, 2005). It is true that weak peasant land rights may affect their decisions over, and capabilities in making, the optimal use of the land. And weak land rights and poorly developed and enforceable laws and regulations contribute to the lack of power of the peasants to defend their interests and participate in policy-making that concerns their livelihoods. But given that policy implementation on the ground remains ineffective and with the augmentation of land fragmentation, it is more important to identify what constitutes or conditions peasants' land rights for sustainable land management and governance. In other words, there is a need to garner social capital and other assets to assist the smallholders in developing appropriate social arrangements to reforming and revitalizing the existing land tenure system, which is a more realistic approach based on local conditions. It is from this angle that the thesis of this book is developed with regard to the conception of pro-poor land tenure suitable for a broader context – not only China.

Pro-poor land tenure should help unravel the unintended consequences of current conventional approaches in terms of land titling and registration of either individual or collective land rights (see Meinzen-Dick et al., 2008). It starts with an understanding of how Chinese society has evolved. Questions as to how peasants' traditional cultures have changed in relation to land, how the logic of political culture has shifted, and how the state and peasants have colluded in political movements need to be addressed. As Zhang (2004) points out, the land reform itself as reflected in the process of collectivization, the formation of People's Communes, the Four Clean-ups Movement, de-collectivization and so forth, were not what peasants themselves had expected or would have chosen. Rather, they were in part imposed by the state.

As the current hybrid land tenure system is far from being a people-centred institution that constructs land relations and land use and management, it is important to explore the changing contexts, relationships and rights to land and examine the changing relationships between land and poverty and how people cope with rural–urban change. The links between land rights, social processes and structures and political and economic organizations deserve further attention. The study of land laws and policies should pay more attention to the issues of social differentiation and inequality. In order to analyse the factors that limit the ability of the poor to pursue their needed rights to sustainable livelihoods, it is vital to address the changing role of land in peasant livelihoods and local social and political relations, which can reveal more practical ways of dealing with poverty and power (see DFID, 2007; He, 2010).

Considering China's development challenges due to population growth and natural resource depletion in the face of industrial and urban development (see Tawney, 1966; Fei, 2006), policy-makers need to put forward a strategic plan for subsistence agriculture and national food grain self-sufficiency on the one hand, and the commercialization of agriculture, industrialization and urbanization on the other (see Pieke, 2005). Subsequently, a pro-poor land tenure system needs to be designed and developed to ensure the sustainable livelihood needs of the peasantry.

However, developing pro-poor land tenure systems is a daunting task as it is embedded within the current context of the HRS and village governance. The loosening of intra-community relations has certainly affected collectively-organized economic activities. This also indicates that the current village collective can hardly act as a genuine entity representing the interests of its constituencies. Thus, the role of the HRS in facilitating market-oriented approaches to land tenure reform and sustainable rural development cannot be overestimated. It serves much more the interests of the local government rather than those of the peasants. Moreover, the limited progress made in village elections has few effects on the empowerment of the poor in the face of the political monopoly of the village collective and local government. The manipulation of the elections by local elites has failed the village collective in providing a significant counterweight to officialdom. A lack of internal conditions such as democratic rules, procedures and capacity of the peasants are hindering village governance processes (Xu, 2003; Lee and Selden, 2007; Van Rooij, 2007; Zhao, 2008a). Although the current institutional framework provides the space for institutional innovation such as the creation of peasant economic cooperative organizations, water users' associations and so forth, these organizations can hardly exert a major influence on more democratic village governance. Thus, 'a more democratic, or at least accountable, land planning regime in China could potentially provide a way out of the bureaucratic infighting and stalemates that have characterized China's land policies for so long' (Pieke, 2005: 100).

As the Chinese peasants are not empowered to participate in land use and management processes, promoting inclusiveness is critical to the fostering of village-based institutional arrangements for pro-poor land management (World Bank, 2003). However, the issues remain as to whether there is a need to create new institutions or to improve the current institutions that can drastically represent peasants' land rights and benefits in the context of those complex social and political relations. Thus, a practical approach is needed to build the institutional basis of innovative land tenure systems that allow flexibility for the peasants to decide on a particular land use arrangement deemed appropriate for the specific village setting. As such, a particular land tenure system should be developed according to the local needs for sustainable development and good governance, which encompasses sound participatory land use planning, among other institutional arrangements.[9]

In practice, this chapter argues that depending on the varying degrees of land use for a specific plot and for the village as a whole, a single land tenure system may not be sufficient. Rather, a plurality of land tenure systems encompassing both individual and collective landownership should be explored. This process requires the development of land users' incentives for their participation in the full cycle of land use planning and management and its associated institutional development framework by attaching great importance to peasants' livelihoods and natural resource use. It impinges upon peasant voluntary action to form land use and management groups when group tenure is appropriate. These groups would be willing to participate in decision-making to ensure that labour and benefits can be shared equally among themselves. In this way, 'these collectives would be built on very different principles from the failed historical examples, and would also offer an alternative to atomized/individual private enterprises' (Agarwal, 2008: 2). Unless they can be convinced of the principles and practicalities of this approach, the land use planners for pro-poor tenure may not stand a good chance of winning their support. Hence, developing pro-poor land tenure is part of the process of helping the poor and the policy-makers identify and overcome the biophysical, social and political constraints to sustainable livelihoods, land use and village governance.

Developing a better understanding of peasants' pragmatisms towards land use and management is essential to the design of diversified land tenure systems. In this respect, China can learn from other country experiences where communities are encouraged to form their networks to carry out their own land use planning with government support (see GLTN, 2010). Moreover, comprehensive policy measures are needed to widen the focus on land use change, and to provide other economic options and policy incentives for the peasants to choose appropriate land tenure arrangements for the sake of sustainable rural development (see Xie et

---

9   A new area-based planning approach that involves specific plans for land reform at local level may be appropriate in South Africa, for instance. This approach helps determine the nature, location, purpose and target groups for land reform, see Hall (2008).

al., 2005). Developing pro-poor land tenure depends on responsible governance, whose challenge is to improve land administration at all levels of government through more effective communication strategies, transparency and accountability mechanisms, as well as promoting citizens' rights to voice their concerns about land management processes (Hilhorst, 2010).

Conclusion

This chapter illustrates China's development challenges are underpinned by various landed factors such as natural resource constraints and land tenure, poverty, poor village governance as well as the need for improvement in land law and policy that can benefit the poor. It explores a wide range of perspectives to discuss the multi-dimensional complexities of land tenure reform, which do not actually invite simplistic approaches. A caveat of the state-led incremental approach that prioritizes the market-oriented land tenure reform can also be costly and more unpredictable. As seen in other country contexts, it is necessary to effectively coordinate the actors involved in the land reform process through decentralized and demand-driven implementation that breaks down the structural barriers of rural development, in particular exploitative power relations and the concentration of capital and market in a few hands (see Hall, 2008, 2009). In this sense, China is no exception, and thus it is more helpful for policy-makers to foster local initiatives.

Ironically, it is important to note that while developing economies, especially from Africa, treat the Chinese land reform as successful examples to learn from in terms of the hybrid system of HRS and village collective landownership and the role of family farms in poverty reduction and farming efficiency, as discussed above, Chinese mainstream policy has already downplayed the HRS in respect of the pursuit of economies of scale. As experienced in Africa and Asia, small-scale farming can be more efficient and more pro-poor than other alternatives given its advantage of creating employment-intensive conditions (Hunt and Lipton, 2011). Hence, the issue does not rest only with farm size itself, obviously. Neither is it about an overt emphasis on who owns what – be it collective or individual households. The crux of the matter for China is that the current hybrid system is not pro-poor enough.

The need for pro-poor land tenure designs in China as an alternative to revitalizing the vast countryside through peasants' own innovation has not been given ample attention in research and policy. In fact, pro-poor tenure designs do not deviate from but are innovating the existing institutional arrangements. By supporting and fostering local initiatives, pro-poor land tenure can facilitate the ongoing decentralization process aimed at enhancing government accountability by offering new impetus for peasant participation in land policy-making and management processes. And it could revitalize the village relations for the formation of more participative local institutions. It would expand the current debates on landownership by offering more evidence-based local experiences as crucial to informed theoretical construction and policy developments.

Future land laws and policies may need to give more attention to peasant participation in decision-making and actively support their initiatives in determining the types of land rights they need for a specific type of land use and management. Innovative forms of tenure should be pursued, which address the constraints on livelihoods and natural resources (see Augustinus and Lind, 2007). To realize this approach, it is necessary to align peasants with the wider public in promoting their development agenda. First, there is a need to understand local development dynamics that pose both opportunities and constraints to sustainable land governance. Second, peasants' perspectives and cultures should be taken into account in land use planning and policy-making processes to ensure that they are supportive of policy changes. Third, civil societies should be encouraged and empowered to participate in this process and given ample space for advocacy and community support. Essentially, policy-makers and researchers should attach greater importance to the conditions and dynamics of land tenure systems in a given setting and render greater support for pro-poor land tenure designs from sustainable land use, development and governance angles, which go beyond the narrow definitions of land tenure security. One can easily fall into the river if one is staggering on those stones that are too slippery to be touched.

References

Agarwal, Bina. 2008. Collective approaches to land rights, *The Broker*, Issue 8, June 2.

Augustinus, Clarissa and Lind, Erika. 2007. *How to Develop a Pro-poor Land Policy: Process, Guide and Lessons*, Nairobi: United Nations Human Settlements Programme (UN–Habitat).

Bao, Xiaobin. 2009. "Rural ecology and development", in *Rural Economy of China: Analysis and Forecast (2008–2009)*, Beijing: Social Sciences Academic Press (China).

Benjamin, Dwayne and Brandt, Loren. 2002. "Property rights, labour markets, and efficiency in a transition economy: The case of rural China", *Canadian Journal of Economics*, 35 (4), 689–716.

Brandtstadter, Susanne. 2003. "The moral economy of kinship and property in southern China", in Hann, Chris M. and the Property Rights Group (eds), *The Postsocialist Agrarian Question: Property Relation and the Rural Condition*, Munster: Lit, 419–40.

Bromley, Daniel. 2008. "Formalizing property relations in the developing world: The wrong prescription for the wrong malady", *Land Use Policy*, 26, 20–27.

Chen, Xiwen. 2009. "Review of China's agricultural and rural development: Policy changes and rural issues", *China Agricultural Economic Review*, 1 (2), 121–35.

China Governance Information Centre (CCGOV). 2006. "失地农民安置典型地区典型数据" (Data on settlement of landless peasants in representative regions), *Lingdao Juece Xinxi*, Issue 26, July 2006, 30–31.

de Soto, Hernando. 2000. *The Mystery of Capital: Why Capitalism Triumphs in the West and Fails Everywhere Else*, New York: Basic Books.

Deininger, Klaus and Jin, Songqing. 2005. "The potential of land markets in the process of economic development: evidence from China", *Journal of Development Economics*, 78 (1), 241–70.

Department for International Development (DFID). 2007. *Land: Better Access and Secure Rights for Poor People*, London: DFID.

Fei, Xiaotong. 2006. *Earthbound China*, Beijing: Social Sciences Academic Press (China).

Finke, Peter. 2000. *Changing Property Rights Systems in Western Mongolia: Private Herd Ownership and Communal Land Tenure in Bargaining Perspective*, Working Paper 3, Halle: Max Planck Institute for Social Anthropology.

Global Land Tool Network (GLTN). 2010. "Grassroots and Land Tool Development", UN–HABITAT, http://www.gltn.net, accessed 10 October 2010.

Han, Sunsheng. 2010. "Urban expansion in contemporary China: What can we learn from a small town?" *Land Use Policy*, 27, 780–87.

Hann, Chris M. (ed.). 1998. *Property Relations: Renewing the Anthropological Tradition*, Cambridge: Cambridge University Press.

Hann, Chris M. and the Property Rights Group (eds). 2003. *The Postsocialist Agrarian Question: Property Relation and the Rural Condition*, Munster: Lit.

Hall, Ruth. 2008. *State, Market and Community: The Potential and Limits of Participatory Land Reform Planning in South Africa*, Institute for Land, Poverty and Agrarian Studies (PLAAS) Working Paper 7, October 2008.

Hall, Ruth. 2009. "A fresh start for rural development and agrarian reform", *Policy Brief* 29, Institute for Poverty, Land and Agrarian Studies (PLAAS), University of the Western Cape.

He, Xuefeng. 2010. "Wither China's system of land tenure?" *China Left Review*, http://www.chinaleftreview.org/?id=45, accessed 11 October 2010.

Hilhorst, Thea. 2010. "Decentralization, land tenure reforms and local institutional actors: Building partnerships for equitable and sustainable land governance in Africa", *Land Tenure Journal*, 1, 35–59.

Ho, Peter. 2005. *Developmental Dilemmas: Land Reform and Institutional Change in China*, London and New York: Routledge.

Huang, Jikun, Hu, Ruifa, Cao, Jianmin and Rozelle, Scott. 2006. "Non-point source agricultural pollution: Issues and implications", in *Environment, Water Resources and Agricultural Policies: Lessons from China and OECD Countries*, OECD report, 267–72.

Hunt, Diana and Lipton, Michael. 2011. "Green revolutions for Sub-Saharan Africa?", briefing paper of Chatham House, AFP BP 2011/01, London: Chatham House.

Lee, Kwan and Selden, Mark. 2007. "China's durable inequality: Legacies of revolution and the pitfalls of reform", *Japan Focus*, http://www.japanfocus.org/-Yingjie-Guo/3181, accessed 23 August 2008.

Li, Shi and Bai, Nansheng (eds). 2005. *China Human Development Report 2005: Development with Equity*, Beijing: China Translation and Publishing Corporation.

Li, Wenjun, Ali, Saleem, H. and Zhang, Qian. 2007. "Property rights and grassland degradation: A study of the Xilingol pasture, Inner Mongolia, China", *Journal of Environmental Management*, 85, 461–70.

Liao, Xingcheng. 2007. 非均衡发展下的失地农民问题 (Problem of the landless as a consequence of imbalanced development), Beijing: Zhongguo Shehui Chubanshe (China Society Publishing House).

Lund, Christian. 2000. *African Land Tenure: Questioning Basic Assumptions*, London: International Institute for Environment and Development (IIED).

Man, Yanyun. 2010. "土地财政　难题求解" (Solutions to land fiscal issues), *China Reform*, Issue 8, 16–18.

Meinzen-Dick, Ruth, Gregorio, Monica di and Dohrn, Stephan. 2008. "Pro-poor land tenure reform and democratic governance", *Discussion Paper* 3, Oslo Governance Centre, United Nations Development Programme.

Migot-Adhola, Shem E. and Bruce, John (eds). 1994. *Searching for Land Tenure Security in Africa*, Dubuque: Kendall/Hunt Publishing Company.

Myrdal, Gunnar. 1968. *Asian Drama*, New York: Pantheon.

Oi, Jean. 1999. "Two decades of rural reform in China: An overview and assessment", *The China Quarterly*, 159, 616–28.

Otto, Jan M. 2004. "The mystery of legal failure? A critical comparative examination of the potential of legalization of land assets in developing countries for achieving real legal certainty", Research Proposal, Leiden: Van Vollenhoven Institute for Law, Governance and Development.

Pieke, Frank. 2005. "The politics of rural land use planning", in Peter Ho (ed.), *Developmental Dilemmas: Land Reform and Institutional Change in China*, London and New York: Routledge.

Qiao, Fangbin. 1997. *Property Rights and Forest Land Use in Southern China*, unpublished Master's thesis, Beijing: Chinese Academy of Agricultural Sciences.

Rozelle, Scott, Brandt, Loren, Li, Guo and Huang, Jikun. 2005. "Land tenure in China: Facts, fictions and issues", in Peter Ho (ed.), *Developmental Dilemmas: Land Reform and Institutional Change in China*, London and New York: Routledge.

Schwarzwalder, Brian et al. 2001. "An update on rural land tenure reform in China: Analysis and recommendations based on a 17-province survey", http://nationalaglawcenter.org/assets/bibarticles/schwarzwalderetal_china.pdf, accessed 20 October 2011.

Sheng, Laiyun and Bai, Xianhong. 2009. "Building a well-off countryside", in *Rural Economy of China: Analysis and Forecast (2008–2009)*, Beijing: Social Sciences Academic Press (China).

Sturgeon, Janet. 2004. "Post-socialist property rights for Akha in China: What is at stake", *Conservation and Society*, 2 (1), 137–61.

Tan, Minghong, Li, Xiubin, Xie, Hui and Lu, Changhe. 2005. "Urban land expansion and arable land loss in China – a case study of Beijing-Tianjin-Hebei region", *Land Use Policy*, 22, 187–96.

Tawney, Richard Henry. 1966. *Land and Labour in China*, New York: M.E. Sharpe.

Todaro, Michael P. 2000. *Economic Development*, 7th edition, Harlow: Addison–Wesley.

Tong Sarah Y. and Chen, Gang. 2008. "China's land policy reform: An update", EAI Background Brief, No. 419, http://www.eai.nus.edu.sg/BB419.pdf, accessed 21 October 2011.

Ubink, Janine, Hoekema, Andre and Assies, Willem (eds). 2009. *Legalizing Land Rights: Local Practices, State Responses and Tenure Security in Africa, Asia and Latin America*, Leiden: Leiden University Press.

Van Rooij, Benjamin. 2007. "The return of the landlord: Chinese land acquisition conflicts as illustrated by peri-urban Kunming", *Journal of Legal Pluralism*, 55, 211–44.

Wang, Weiguo. 2005. "Land use rights: Legal perspectives and pitfalls for land reform", in Ho, Peter (ed.), *Developmental Dilemmas: Land Reform and Institutional Change in China*, London and New York: Routledge.

Wang, Yan. 2011. "Agriculture, food security and rural development", *Proceedings (Volume Two) of the Policy Symposium on Economic Transformation and Poverty Reduction: How it Happened in China; Helping it Happen in Africa*, organized by the China-DAC Study Group and International Poverty Reduction Centre in China, 37–66, unpublished report.

Wang, Zhiyuan. 2009. "农村土地制度改革及今后的完善思路" (Rural land institutional reform and suggestions on its improvement), *Beijing Social Sciences*, Issue 5.

Wang, Zhuo and Xu, Bing. 1996. 中国农村土地产权制度论 (Debates on China's Rural Land Property Rights Debates), Beijing: Economic Management Press.

World Bank. 2003. *Sustainable Development in a Dynamic World: Transforming Institutions, Growth, and Quality of Life*, Washington, DC: The World Bank.

World Bank. 2007. *World Development Report 2008: Agriculture for Development*, Washington, DC: The World Bank.

Xie, Yichuan, Yu Mei, Tian, Guangjin and Xing Xuerong. 2005. "Socio-economic driving forces of arable land conversion: A case study of Wuxian City, China", *Global Environmental Change*, 15, 238–52.

Xu, Yong. 2003. 乡村治理与中国政治 (Rural governance and Chinese politics), Beijing: China Social Science Press.

Yang, Li. 2007. "Rangeland governance: How to improve the household responsibility system?", *China Rural Economy*, 276, 62–7.

Zhang, Xiaojun. 2004. "Land reform in Yang Village: Symbolic capital and the determination of class status", *Modern China*, 30 (1), 3–45.

Zhao, Yongjun. 2008a. "China's new development agenda: Democracy Beijing-style", *The Broker*, 6, 4–6.

Zhao, Yongjun. 2008b. "Land rights in China: Promised land", *The Broker*, 7, 7–9.

Zhao, Yongjun. 2010. *China's Rural Development Challenges: Land Tenure Reform and Local Institutional Experimentation*, Groningen: Centre for Development Studies, University of Groningen.

Zheng, Yisheng (ed.). 2004. *China Environment and Development Review*, Beijing: Social Sciences Documentation Publishing House.

Zhou, Qiren. 2007. "试办土地交易所的构想: 对成都重庆城乡综合配套改革试验区的一个建议"(Assumption on land transaction markets: A suggestion on comprehensive urban–rural reform in pilot areas of Chengdu and Chongqing), *Nanfang Zhoumo*, 11 October 2007.

Zhu, Keliang and Prosterman, Roy. 2006. "From land rights to economic boom", *China Business Review*, http://www.chinabusinessreview.com, accessed 15 December 2009.

Zhu, Keliang, Prosterman, Roy, Ye, Jianping, et al. 2006. "The rural land question in China: Analysis and recommendations based on a seventeen-province survey", *Journal of International Law and Politics*, 38 (4), 762–839.

Chapter 5

The Withering Household Responsibility System and Sustainable Natural Resource Management in North China

Introduction

This chapter is an empirical study of the pitfalls of current land tenure systems characterized by the Household Responsibility System (HRS) and its associated rural economic reform measures in sustainable natural resources (land, water, forest, grassland) and the overall challenges of rural poverty and governance in an attempt to contribute to a renewed understanding of its social and political complexities. Guyuan County, a nationally designated poverty-stricken county in Hebei Province is selected as a typical sample of these challenges in North China.

This case study is primarily based on the fieldwork conducted in Guyuan County in 2008 with the support of other related research until 2011. The research methods were mainly qualitative. It was extremely difficult to conduct household surveys, and formally published materials were of largely no avail. Participant observation was used as the main tool to gain access to information on the livelihoods and land use practices of the interviewees, some of whom were reluctant to speak out given the difficulty of the research topic. Nonetheless, it was possible to gather the views of 30 informants from county and township governments and local peasants in eight villages of slightly different economic and natural conditions. The field results were supplemented by government published and unpublished reports and policy documents.

Overall, China's northern regions have seen chronic poverty and acute environmental degradation over the last decades despite a remarkable level of economic development. Land desertification is a primary obstacle for the Chinese government in achieving its targets for faster and sustainable growth. For instance, dust storms are affecting the capital city Beijing and other major cities especially in the Northern Plain, contributing to the deterioration of air quality. It calls into question the sustainable use of farmland and grassland in the agro-pastoral zone under semi-humid and semi-arid conditions. Hence, the management of land desertification has become a major environmental agenda of the government (Wang et al., 2005). This region is also among the most noticeable ones that have experienced a sharp reduction of cultivated land, which poses a major challenge to the livelihoods of the vast poor peasantry (Lin and Ho, 2003).

Greening the region through ecological construction prominently embedded in the conversion of farmland to forest or the 'Green for Grain' programme is a major policy initiative of the central government to drastically increase vegetation coverage (Jiang, 2006). However, this effort is severely undercut by the adverse effects of economic development on natural resources, as most reform policies have accelerated rural land degradation (Williams, 1996; Muldavin, 1997; Sanders, 1999). Moreover, the greening programmes backfired, as relevant practices had not been in full accordance with local ecological processes. By focusing on greening or equating ecological construction with intensive land-use practices, such a pursuit of short-term gains has met unintended social and political consequences as well (Jiang, 2006).

Furthermore, the organization of ecological construction programmes is predicated on the HRS. The latter allows farmland to be relocated to households on an equal footing. In ecologically fragile regions, this leads to the fencing or enclosure of grassland for grass rehabilitation and tree planting on a basis of individual households taking care of the vegetation to ensure more effective and sustainable land use (Jiang, 2006). Due to the assumption that peasants or pastoralists would have ample space to exercise their rights and gain direct benefits, this approach only met with partial success in the early stages of the market reform, as did agricultural development. Overall, by the mid-1980s, China's total agricultural output grew by no less than 7.4 per cent per year (Huang, 1998). Chinese peasants had enjoyed greater freedom to sell surpluses after fulfilling obligatory grain quotas as compared with the commune era. However, after 1985, China's agricultural growth slowed. Rural environmental degradation and depletion of natural resources made many poor people fall back into poverty (Li et al., 2004).

To capitalize on the advantages of the current HRS, some scholars argue that China would need a more individualistic institution that facilitates the development of tradable land rights or a rural land market under the rule of law (Lai, 1995; Cai, 2003; Chin, 2005; also see Ho, 2005; Szirmai, 2005). By contrast, Bramall (2004) contends that one should not overestimate the role of HRS in the Chinese agriculture. Rather, government intervention, technological advancement and natural conditions have played a more important role. The current small-size household farming system has caused major problems. These include fragmentation of land, land lost to paths and boundaries and conflict over access to irrigation systems among village groups. Moreover, it makes large-scale agricultural production extremely difficult. Access to land has not been the basis for China's agricultural prosperity. Land is valuable because of price support for agriculture rather than the greater efficiency of small-scale farms.

Transforming the HRS into more individualistic land tenure may not provide a viable solution to unsustainable natural resource management. In a similar vein, Hu (1997: 175) points out that the current land tenure system has encouraged short-sighted decisions and irresponsible use of land resources. Peasants pursue immediate and short-term gains at the cost of environmental sustainability to a certain degree, which is exacerbated by land fragmentation. The latter hampers irrigation and drainage and leads to the degradation of the agro-ecological environment. Local governments do not function effectively in organizing agricultural production or overall rural development, due to a lack of resources and democratic governance. On the one hand, the lack of resources and sound governance has hindered their role in sustainable rural development. On the other hand, slow agricultural development and insufficient fiscal transfers from higher-level governments have generated insufficient resources for local governments to deliver basic rural services. Moreover, the Chinese peasantry, to a large extent, has not been organized in a way that their land can be better utilized and managed to their own advantage.

In semi-arid areas such as the Guyuan County, the wide application of the HRS by the local government to avert the 'tragedy of the commons' (see Hardin, 1968) has not come to grips with complex ecological, social and political conditions. This chapter sheds light on the current debates as given above and challenges any preconceived model of land tenure with an attempt to explicate its linkages with natural resource governance and rural development. It shows how the HRS has contributed to the fragmentation of rural social and political relations. It also explores the underlying institutional constraints especially concerning the major pitfalls of the HRS in governing the use of land resources by the poor whose livelihoods have become more vulnerable to the government management schemes on these resources – farmland, grassland and forestland. It manifests the linkages between peasant livelihoods and land-induced conflicts among different actors over resource utilization. It suggests the need for policy redressing to allow for peasants' choice over the design of more appropriate land tenure systems that contribute to sustainable natural resource use and management and poverty alleviation in China's dryland.

Ecological and Development Constraints

The mountainous Guyuan County has a total area of 3,654 square kilometres and a population of 230,000, among whom 196,000 are peasants (see Figure 5.1). Agriculture and animal husbandry characterize its economic landscape accounting for 70 per cent of the total county GDP. With an income of RMB 1,480 (US$231.61 as at August 2011) per capita, its severity of poverty remains a major challenge for the county government to tackle (Bi, 2010: 1). Moreover, with an average of 0.43 ha of arable land per capita, there has been limited progress made, especially in the promotion of large-scale vegetable farming since 1998 (Guyuan Poverty Alleviation Office, 2007).

Guyuan also falls within the ecologically-strategic region under the *Three North Shelterbelt Programme* aimed at preventing China's semi-arid and arid land from further degradation. The Chinese government started this programme in 1978 with a budget of 40 billion Yuan over 70 years to create 35 million hectares of man-made forests. Perhaps as the largest government project to re-engineer the rural landscape (Jiang, 2006: 1913), the progamme is also envisaged to prevent sand storms from entering the inner regions such as Beijing. With an average precipitation of 392.3 mm (Wang et al., 2005: 2403), the county like others in the region faces severe water shortages. The North China region as a whole produces almost 25 per cent of China's total agricultural output, although it has at its disposal only 5 per cent of the country's water resources. Irrigation is extensively used in agriculture (Kahrl et al., 2005: 13). A large part of its soil is covered with sparse vegetation as a result of salinization and alkalization, although large tracts of grassland and forests are spread out in certain parts (Wang et al., 2005).

Figure 5.1   Guyuan County, Hebei Province, China

Furthermore, Guyuan is highly prone to natural disasters, such as drought which affects an average of 30 per cent of its farmland. Economic development has exacerbated this situation, as the loss of farmland, grassland and forests are rising. Accordingly, the natural resource base especially soil fertility and groundwater level has declined dramatically. Hence, unsustainable natural resources use and management sets further constraints to agricultural development. In addition, inadequate access to public infrastructure and technical services has prevented many peasants from farming and marketing their produce in an efficient manner (Guyuan County Government, 2003: 97).

Since 1998, the county government has spearheaded the development of agribusiness enterprises on the assumption that this trajectory would enable different villages to develop their economies of scales. Animal husbandry and vegetable farming have been promoted as the most important enterprises for poverty reduction. It has become a well-known region for supplying milk, beef and vegetables to other parts of the country, especially the North. In 2007, for example, 80 per cent of the villages or 40 per cent of the population were involved in vegetable farming, using 15 per cent of the arable land (Guyuan Poverty Alleviation Office, 2007).

By contrast, before 1998 cash crop farming in the county was minimal. Traditional crops such as oats and flax were widely planted for both domestic use and the market, with tiny profits gained. Paradoxically, these crops are more drought-resistant than vegetables. The introduction of the latter was assumed to be a quick fix to prolonged poverty, yet the extent to which poverty has been reduced is limited. While peasants shifted their traditional farming to the 'modern' forms, the majority of them have not benefited from this switch. Natural disasters, water shortage and a lack of collective organization of farming constrain peasants' efforts in maximizing farming efficiency and market access. Both the county government and the peasants have voiced their concerns about the sustainability of the current farming methods in the light of these challenges. Above all, peasants' lack of information on the market, lack of choice over farming, and lack of off-farming employment opportunities, has further constrained their pursuit for livelihood enhancement.

Changing Land Relations: Mutual Help, Conflicts and Cooperatives

Understanding the history of land tenure reform is important to the analysis of changing land relations which are central to rural development and the governance of natural resources and village affairs. In fact, land reform in Guyuan differed little from the rest of the country, but is exemplary of the mainstream policy trajectories and approaches with its particular characteristics of ecological, poverty and institutional constraints. Prior to the founding of the People's Republic by the Communist Party in 1949, most agricultural land in this region was owned by landlords, rich peasants and merchants, whereas ordinary peasants owned little or no land at all. Only a small number of poor peasants maintained their smallholder status based on many years of hard work and savings. There was a high level of social and economic inequality as a result of uneven landownership. The majority of the peasants rented land from the landlords and rich peasants. Land transactions took the forms of leasing, purchasing and mortgaging.

In the aftermath of the land revolution led by the Communist Party calling for the abolition of the exploitative feudal relations, many landlords' landed properties were confiscated and redistributed to the poor peasants. This victory faced an immediate challenge for agricultural development, since it was a drastic process of severing the old productive relations. The peasants with redistributed land could hardly cope with the shortage of labour, livestock and machinery, which were all essential to efficient farming. Peasant cooperation became a necessary institution to deal with these problems. After 1950, the Party's policy promoted the establishment of temporary and year-round mutual help groups based on voluntary principles. These groups played an important role in offsetting the shortage of human and technical capital through the exchange of labour, livestock and machinery. During this period the number of the mutual help groups increased substantially. In 1950, only 2.8 per cent of the households were involved; but by 1954, this number reached 81.7 per cent (Guyuan County Government, 2003: 197).

The consolidation of peasant mutual help groups led to the establishment of more organized peasant production organizations with strong government intervention. This transformation went through three stages. At the first stage, in 1952, primary agricultural cooperatives were piloted and rolled out to the whole county. Individual households remained as the landowners, but also as cooperative members to receive benefits based on their labour contribution. Land use, management and agricultural production were all arranged by these organizations. At the second stage started in 1956, the primary cooperatives were transformed into more advanced agricultural cooperatives. Collective landownership replaced the old private ownership. All peasant households automatically became members of the cooperatives, which arranged farming and distributed production materials to the members. At the third stage, these advanced cooperatives were transformed into communes – a higher-level peasant organization equivalent to the current institution of township, under which collective ownership of landed resources displayed more bureaucratically centrally-controlled mechanisms. Subsequently, the commune quickly showed its ineffectiveness in farming organization and rural development. The local government attempted to improve its efficiency, but failed to provide ample incentives to the members to stimulate agricultural production. The Cultural Revolution (1966–1976), characterized by fierce political struggles, contributed to the destruction of social and economic relations at all levels, which obstructed further improvement of the commune (Guyuan County Government, 2003).

The HRS was first introduced in 1979 in pilot villages with great difficulties. Similar to the realization of the previous policies on the models of cooperatives and communes, its adoption was sanctioned through stringent administrative measures, without full peasant consent. Land, labour, livestock and machinery were allocated to individual households who were given the responsibility to meet production and other economic quota and taxes set by the local government. On the assumption that the HRS would provide the peasants with more incentives to cultivate their land, its outcome has not been prominent due to poverty and natural resources degradation. Moreover, the HRS from the very beginning cultivated the seeds of inequality, as large farms were leased to the so-called capable households. During the 1970s and 1980s, these farms were run with a huge loss of profits and property due to mismanagement and weak governance. In 1993, the whole county followed the call of the central government to stabilize and improve farmland contracting relations by granting 30 years of land use rights to the households. And in 1997, a second round of farmland leasing was carried out with a view to clarifying and documenting land contracting rights and improving land tenure security. The latter was assumed to be vital in stabilizing land relations and facilitating land use rights transfers among the households, as some of them would prefer leasing their land to others while undertaking off-farm employment (Guyuan County Government, 2003). In short, the egalitarian principle and practice concerning agriculture gradually receded with the introduction of the HRS whose alignment with state policies on agricultural output to be met by households does not always suit peasants' needs for livelihoods and social services (Chang, 1994).

The practice of farmland leasing also triggered land conflicts among contractors, village collectives and local government after 1997. Affected peasants lack the power to hold the local government accountable or to negotiate terms of conditions with it. Village collectives are sometimes accused of mishandling of land allocations. Those households with close relationships with the village leaders sometimes receive more, and better quality, land than the rest and may not fully comply with their contractual terms. Quite often, the land of those who have migrated to cities is intentionally kept and redistributed to others. In the event of the migrants returning to their homes when they were unable to find a permanent stay in the cities, they found that their land had been leased to other households. This is the most critical factor for disputes and conflicts between the returned peasant migrants and village leaders.[1]

Obviously, how to manage appropriate land use for the benefit of the poor presents a daunting challenge for local governance. This challenge also exhibits the weakness of the HRS in securing peasants' land use rights. This historical account shows that none of the land tenure regimes have worked effectively, to a certain extent. Rather, they have undermined the power of the poor and led to rising social inequality and conflicts. Against this backdrop, current agricultural development policy and practice has attempted to call for a shift from individualistic to more cooperative approaches as embodied in the institutionalization of specialized farmers' cooperatives, in accordance with the promulgation of the *2007 Law of the People's Republic of China on Specialized Farmers' Cooperatives (SFCs)*.

SFCs are assumed to be instrumental in facilitating individual peasant households' forays into the market and thus contributing to the development

---

1 Rural outmigration may not contribute to rural development to a large extent given the fact that enormous urban employment creation is needed to accommodate the migrants. However, it is never an easy task. For rural development to take off substantially, 75 per cent of the peasants have to leave the countryside, and this is almost unrealistic (see Kahrl et al., 2005).

of more specialized agricultural enterprises to greatly enhance agricultural productivity and poverty alleviation effects. It is deemed as a necessary outcome of the modified system of rural economic governance structure in general and the HRS in particular given the latter's constraints to economies of scales. Using the HRS as the basis of an economic unit, the SFCs are meant to be organized in alignment with peasants' willingness to participate and manage in order to maximize their roles and to ensure their rights, duties and benefits. Local government and village collectives normally play an essential role in supporting the SFCs in terms of organization and technical support, and village experts and local enterprises are the 'dragon heads' in charge of operations. Since 2007, 24 cooperatives that involve 1.3 per cent of the peasant households have been founded, which cover a wide span of varied enterprises related to the produce of vegetables, corns, edible fungus, milk and chicken, and so on (Zhou, 2010).

However, the further development of these organizations is facing various constraints in terms of lack of scale and starting capital, low profile or lack of interest among peasants in becoming members, lack of organizational management capacity, especially concerning financial transparency and accountability, among others. As a result, they remain as 'empty shells', or just serving the purpose of the local government policy agenda as a matter of formality. For the county government, overcoming these constraints lies in essentially offering the peasants the needed economic incentives in institutional development to support truly peasant-centred organizations. Moreover, as agricultural enterprises concern the interests of most peasants, facilitating more specialized enterprise development through land use rights transfers is deemed to be the key to more scaled production of the raw materials needed and the formation of cooperatives. By doing so, the county government followed the model of 'dragon head enterprises + cooperatives + resource-base + peasants' as tested in many parts of China with limited success to form various cooperative relationships among these stakeholders. By doing so, it is assumed that peasant vulnerabilities to market risks can be overcome (Zhou, 2010). However, it is far too early to assess the sustainability of this model given the lack of peasant interest and mistrust in these enterprises with regard to elite capture and lack of empirical research (see Chapter 6).

Overall, it can be seen that the role of the HRS has largely been downplayed in the county government's pursuit of fast rural development outcomes, as elsewhere in China. To a large degree, it is even seen as an obstacle to facilitating the formation of cooperatives as well as realizing economies of scale given the nature of land fragmentation. Notwithstanding the short-term economic benefits to the members as claimed by the local government propaganda, the cooperatives established in the current format trigger stakeholder conflicts, as power is vested in those leading agents or the so-called capable experts.

Unsustainable Farmland Use and Peasant Contestations

Given chronic poverty, rural development for the local peasants has a strong bearing on the utilization of the available resources. Their attachment to land, pasture, forests and other resources embodies their pragmatic values towards their livelihoods. Yet, peasant relations have become less dynamic than the those found in the People's Commune period in the 1960s and those found in traditional ethnic minority groups. With the market-oriented policies penetrating their communities, individual interests override the mechanisms of collective choice and decision-making on the use of the resources for the benefit of all. This is inseparable from the way the natural resources including the land are utilized, to a large extent unsustainably.

In China, between 1995 and 2001, the production of vegetables nearly doubled (Lichtenberg and Ding, 2008). In Guyuan, during the course of ecological rehabilitation as mentioned earlier, local and even regional agriculture has undergone tremendous structural changes. As a result, widespread vegetable farming contributes to the county policy on agricultural production and thus constitutes a primary source of income for most households. However, farmland fragmentation has reduced the efficiency of farming as evident in vegetable farming, which is akin to 'digging the soil and land without caring about the kind of resources they will leave for the future' (Hu, 1997). Although soil conditions are favourable, this sector is facing the challenges of acute water scarcity and depletion as it involves a high-level consumption of water in addition to the use of chemical fertilizers and seeds. A local saying 'selling vegetables is the same as selling water' sends a clear warning in this regard. Not only has water become a bottleneck for this industry, but also has contributed to the aggravation of desertification and the destruction of vegetation planted under the 'Green for Grain' programme.

Where there is a lack of technical and funding services provided by the local government, the vegetable growers are vulnerable to, and incapable of effectively dealing with, various natural and economic risks. The varieties of vegetables grown are monotonous across the county, and this unavoidably leads to competition over sales and marketing outlets. Only those households with relatively larger parcels of land manage to gain reasonable profits. Since they have no other ways to sell their produce than relying on the middlemen or the so-called specialized cooperatives to collect it, quite often they are in a weak bargaining position over the prices offered.

Village relations are now much more complex than the past when inter-household cohesion and mutual help played an essential role in organizing the peasants.

> Everyone is helping himself. We do not know about the future – we just try to make ends meet anyway. Nobody will help us. (Villager)

This kind of remark is commonly found in the villages, and is also shared by local government officials. Despite the partial success of HRS in poverty reduction, it displays an increasing weakness in uniting the poor as land becomes fragmented. The peasants have less space for the social organization of agricultural production. And when the village collective mostly represents the interests of the local state, the role of the peasants in voicing their concerns over land use and agricultural development becomes minimal. All the informants held the view that they were not sure what land relations in their villages existed, as they worked on the land on their own. They were puzzled by the constant changes in government policies as a whole as to whether their land rights would be altered by the local government at any time. Facing uncertainty over land use and other associated rights and a lack of public support, many informants viewed the Chinese peasantry as the most vulnerable group in the country, and they cast doubt on whether peasant collective relations can be revitalized; some were even reluctant to see the diminishing role of the HRS in securing their livelihoods.

The fragmentation of farmland use caused largely by the introduction of the HRS underpins complex peasant–local state relations as both have different or even conflicting interests in the land. As the value of the land increases with local government gaining increasing control over it, local peasants see their land as the last resort to maintain their livelihoods. The fieldwork found the following case in point characteristic of this argument.

The Pigpens Case

The national policy on farmland protection places strict conditions on farmland use and prohibits its conversion into non-agricultural uses unless this is done under state expropriation. Accordingly, the local government set the mandates for the village collective to demolish all buildings such as pigpens in the fields, and peasants' ignorance of this call would lead to forced demolition of these properties. This loss would incur a threat to their livelihoods, as a large proportion of their incomes depended on livestock rearing. However, when the peasants built their houses years ago, they were not informed of whether it would contravene any government policies. In fact, their ideas were even approved by the village collective. As these households of small numbers only occupy tiny plots unsuitable for cropping, this act literally has nothing to do with farmland conversion. Seeking the support of the village committees was futile. As many of them were women, children and the elderly left behind in the village, they could not form a unified force against possible land evictions that paved way for new land development projects.

In X Village,[2] a household led by a 50-year-old woman with three children has farmland of slightly above 17 *mu*, equivalent to just 1.1 ha, which represents quite a big household in terms of land occupancy. In fact, part of the land is rented

---

2   Pseudonyms are used for villages illustrated throughout the book.

from one of her brothers and others. This practice is quite common in rural China as many peasants seek temporary employment in the cities and sub-lease land to others from the same village or even other regions. In this respect, her husband works in a city in the province. Given her relatively large land size, in peak farming seasons, she needs to hire other peasants to help her with cropping.

Vegetable growing constitutes the largest cropping pattern, whereas livestock rearing serves as the major source of income. She rears more than 50 pigs, which contributes to an annual total income of RMB 30,000–40,000. However, her net household income remains minimal after the deduction of the rising costs of fertilizers, and other materials. She seems quite aware of what is going on in the village but is reluctant to tell more about the underlying issues surrounding the land, livelihoods and village affairs. To her, there are no other alternatives to growing vegetables and rearing pigs, even though she is fully aware of the environmental negativity, market risks and vulnerability to demolition of the pigpens. She reflects that most peasants feel powerless in their demand for sustainable use of the land in the face of land tenure insecurity. In the case of land acquisitions that have been occurring in the surrounding villages, there is little they can do to overturn the terms and conditions offered; and given the average compensation fee of RMB 30,000–40,000 per *mu*, it would be meaningless for them to give their land away as they see land values rising. The foreseeable demolition of her pigpens would be a heavy blow to her livelihood, which could be aggravated by farmland acquisition at a later stage by the local government given the close distance between the village and the town. When their discontents reach a tipping point, many are reluctant to seek legal or political action, since previous cases of petitions were turned down by the local government. She seems to hint that land privatization would be ideal for them to gain power given the fact that land relations have already been fragmented, but she does not know what better land relations can be forged.

In fact, land use planning and management has never been an easy task for the local government. In the absence of sizeable business enterprises and foreign investments, the county has the mandate to create a favourable business environment by providing the necessary basic infrastructure and land. Thus, the acquisition and consolidation of land has become a crucial step towards this end. At the same time, the local government has to restrict the use of farmland for 'non-agricultural' purposes such as the construction of pigpens by the peasants in order to strike the overall balance between farmland preservation and conversion.

Moreover, the county government line agencies are trapped on the path towards economies of scale in agricultural production, making the maximization of land use and land acquisition unavoidably difficult for them. Some seemed to agree that the county had no other options but to introduce the 'dragon head' agribusiness to take the lead in organizing scaled production. They complained about the peasants' backwardness of ideas, knowledge and skills in adjusting to the demands of the market economy. However, for the peasants, the lack of secure land rights and the mechanisms for transparent and effective partnership with agribusinesses could only make them cast doubt on new institutional arrangements. As the local government has a mandate to push further ahead with economic development, its conflicting interests with the peasants are on the rise as the pigpen case unveils.

Impractical Policies and Rising Conflicts

Reflecting on those SFCs schemes experimented in southern China, the county line agencies staff pointed out that they should allow the peasants to become land shareholders and benefit from agribusinesses.[3] Thus, the village collective should play a stronger role in uniting the peasants and assuring them better economic returns on agricultural production. However, the current policy on these organizations requires the registration of a substantial amount of capital, which precludes the participation of the majority of smallholders. Moreover, government officials also realized the fragmented rural social structure that hampers effective institutional innovation. Nearly 50 per cent of them held the view that the trend of the rural economic reform should reverse the HRS into more genuine collective land use and management as a way to revitalize the countryside and help the peasants cope with their varied vulnerabilities associated with land use. However, how to convince the peasants and what incentives to offer them remains a critical challenge.

Furthermore, peasant–cadre relations have become more complex. Changes in the central government policy that favours agriculture, on the one hand, have created more incentives for the peasants to care about their land; on the other hand, they also spur local conflicts. Land-induced peasant–cadre conflicts have become a thorny matter. A recurring example is the tension between returning peasant migrants and local cadres over land reallocation. Village collectives and township governments are often accused of purposely reallocating the out-migrants' land to other households in the name of efficient care of the land to avoid its waste as well as accommodating demographic changes. In this case, Guyuan resembles the rest of the country, where large numbers of impoverished rural people have become migrants to join the 'floating population' of 200 million seeking temporary work in the cities (see Zhang, 2001; Solinger, 2002). To the township government, the land left by these migrants should be redistributed to accommodate the needs of other groups. Paradoxically, this practice is no longer permitted by law, which aims to ensure land tenure security and explicitly calls for an end to land readjustments.[4] As one member of a local township government remarked,

---

3   However, shareholding cooperatives have their limitations, since the mechanisms for mutual supervision and self-restraints are often inadequate, representation of the vulnerable poor is not strong, and the interests of capital may predominate. See the detailed discussions in Chapter 6.

4   Article 27 of the 2002 Rural Land Contracting Law states that land adjustment is prohibited during the contract period. However, Article 28 states that land adjustment should be done on the land returned by the contracted households to accommodate newly increased households. Thus, it leaves space for ambiguous local interpretation, as it is hard to tell which land belongs to the returned migrants.

Of 26 administrative villages in Guyuan, 18 villages or at least 80 per cent of the cases have involved peasant petitions over land use issues. The current land law and policy are much to blame, as they do not really take into realistic account the local conditions. And it is sometimes contradictory in terms. For instance, it enshrines women's land rights. But if land readjustments are not allowed, how can we give land to the women who marry men in our villages? Besides this, the recent agricultural policies featuring economic incentives for the farmers have actually extended an invitation to the migrants who want to return to farming. When they migrated to cities, they left their land idle and let us manage it. We then leased it to others who could farm the land. How can we return this land to them when it is in others' hands? ... If you want to study rural China, study the rising rural conflicts which are of prime concern of the government. (Interview with local township official, July 2008)

These conflicts underlie the peasants' growing concerns about their rights, livelihoods and ineffective policies in the context of widening social inequality between different groups. As a result, land readjustments may privilege some while marginalizing others due to the practice of personalism, clientelism and networking tactics carried out by the dominant group (see Nonini, 2008). Moreover, peasants' struggles for land are sporadic and seldom organized systematically in terms of the creation of effective groups that can maximize their influence and collective force, which explains the weakness of the HRS in realizing people's potential for social cooperation.

Paradoxically, 90 per cent of the respondents indicated that they would rather keep the HRS, as they did not trust the village collective. The rest either preferred the old commune system or simply had no preferences. Some even expressed their inclination towards land privatization. But all of them expressed their concerns about the absence of viable property relations for sustainable land use and management and poverty alleviation as a whole.

Ineffective Grassland Enclosure Institutions

Following the central government's appeal for scientific development with a strong focus on equity, social harmony and sustainable development, Guyuan has a mandate to implement the national and provincial ecological construction plan. The optimized use of its fragile natural resources such as grassland has thus become a primary goal of the county government. Grassland enclosure and animal husbandry prohibition is strongly favoured in relevant policies deemed as 'scientific' in tackling the declining carrying capacity of the grassland. By doing so, the local government has a strong determination to ban the traditional method of grazing officially claimed to be disorganized and attributable to the tragedy of the commons. As such, this move chants a major accord with the 2002 Grassland Law of China that sets a strong mandate for local government to properly preserve the grassland. Under this law grassland ownership rests with the state which assigns use rights to the village collective. The latter is allowed to lease the land to individual households. In particular, Article 33 states the following:

> Contractors for grassland management shall make rational use of the grasslands, and they may not exceed the stock-carrying capacity verified by the competent administrative department for grasslands; and they shall take such measures as growing and reserving forage grass and fodder … in order to keep the balance between grass yield and the number of livestock raised. (Government of China, 2002: 7)

This stipulation reflects the government's arbitrary approach to grassland management in applying the carrying capacity concept. It marks no difference from the 2002 Rural Land Contracting Law in terms of granting land use rights to individual households. Thus, a large part of the grassland in Guyuan has been partly contracted out to individual households who are required to sign their use rights contracts with the county Agricultural and Animal Husbandry Bureau. In principle, their grazing rights are set against the numbers of livestock to be kept by them. Ideally, the delimitation of household-based grassland is assumed to automatically lead to a reduction in the number of livestock, as the peasant households are held accountable. In a few cases, some parts of the grassland are kept in the hands of the village collective purely for the purpose of nature conservation. This land is fenced off for rehabilitation, as either it has been exploited almost to extinction or it is prone to further degradation. Apart from the contracted and preserved grassland, there is only a small proportion of the land left open for communal grazing.

In practice, however, grassland protection is too costly and difficult to manage for the county government, since the peasants can find ways to cut the fences and enter the prohibited areas. In addition, the grassland contract management has not succeeded in fully registering the peasants, some of whom have not applied for the contract certificate as required by law. Although those households with the certificates are allowed to graze appropriate numbers of livestock, one can hardly tell whether these numbers have been followed. There is a lack of institutional mechanisms for effective monitoring as county and township governments and the village collective do not have clear roles to play and thus do not think that they are the sole responsible institutions. They often blame each other for the lack of more coherent and enforceable policies.

The ineffectiveness of the law in the view of the local government is due to peasants' lack of 'modern knowledge' of livestock rearing and grazing. To avert overstocking, in referring to the proclaimed 'scientific knowledge' derived from the basics of ecology (see Jiang, 2006), the local government made further attempts to introduce new grazing methods to the peasants. However, these methods require improvements in the use of fodder and feed, which are too costly for the peasants to purchase and stock. The peasants have continued to ignore these calls and managed to avoid the inspection of the local bureau staff regarding use of the grassland. Some rehabilitated sites, despite being under a certain level of well protection for some time, have now become degraded again due to peasant invasions. Some large parcels of grassland contracted out were found to be returned to cropping and other uses, which is totally forbidden by law. But in general, with the implementation of the 2003 Decree of Grazing Prohibition, access to grazing has become more and more difficult for the peasants, who complained that their income from farmland cultivation was so limited that they had to rely on grazing to supplement it. The lack of adequate access to grassland further aggravates their fragile livelihoods (Xinhua Net, 2006).

Overall, the continuation of grassland degradation indicates relevant institutional failures which put overt emphasis on administration measures rather than seeking viable options for peasant livelihood enhancement. To the peasants, the traditional method of grazing has certain cost-effective advantages, including flexible management of the livestock. As their own grazing rules are not considered by the government, they do not believe that the exotic method of grassland enclosure is in their best interests. In fact, as in other regions in China, grassland enclosure has contributed to overgrazing and thus land degradation (Jiang, 2006). Since there is not much leeway for peasants to utilize their own contracted grassland of small in size, grazing on the preserved grassland becomes unavoidable. This fieldwork found that 95 per cent of the village respondents knew what was happening, but did not know how to deal with it. The rest simply did not believe that they themselves should be blamed. Above all, they saw livestock rearing as a better way to fight poverty than vegetable farming, but they had to bear increasing risks.

The ineffectiveness of policy instruments indicates that, as in many other parts of the country, grassland preservation programmes are not coupled with appropriate poverty alleviation strategies. Moreover, no community-based land management models based on household tenure are found (see Banks et al., 2003). In Guyuan, grazing remains as the most beneficial means of livelihoods for those with relevant skills. They showed discontent over the county government's grassland reclamation policy which caused an increase in poverty and rising social conflicts in the village (Xinhua Net, 2006). Their views on participation in grassland use and management have not been placed at the top of local government development policy.

Furthermore, grassland management is often compounded by the fuzziness of the heterogeneous and hybrid property relations, where village collectives and even local governments have more power to determine land uses than the peasants, and where lines of responsibility of grassland management are not clearly demarcated among different state and non-state actors (see Hinton, 1990; Yeh, 2004). These factors contribute to poor grassland governance. A prominent example is the Ministry of Agriculture's national circular in 2006 publicizing the misconduct of Guyuan local government staff and land contractors in grassland use.[5] It was reported that the county government and a village committee contracted large tracts of grassland owned by the government and village collective respectively to a local businessman and certain peasant households without prior notice to the villagers, nor were transparent procedures put in place. The local community simply did not know how this was done and only found out later that the land was converted into cropland and other usages. According to the 2002 Grassland Law of China, land contractors must maintain the original status of the land upon transfers. In this case they managed to cover up their purposes by stating that they wanted the land for eco-tourism development and would take care of the land to abide by the law. More strikingly, the two contractors even occupied the land before the local bureaus formally approved their applications (Ministry of Agriculture, 2006). This means that to a certain extent, the current grassland tenure system underpinned by the HRS has favoured the powerful groups in their use of the land for their own benefit rather than reaching the assumed objective of resource use efficiency and sustainability.

Forest Plantation Schemes: Quantity but not Quality Driven

Another illustrative case rests with forest plantation schemes. The county government has made forest plantation of equal importance with grassland preservation to realize its plan of ecological rehabilitation and expansion. Under this plan, half of the county's farmland would be converted to forests by 2010, which would affect more than 30,000 peasant households (Bi, 2010). To a large extent, their measures have been effective in terms of the quantity of trees planted under the 'Green for Grain' programme, for instance. However, peasants are not offered strong incentives in tree planting and do not derive expected benefits from it. For instance, in recent years peasants received only an estimated RMB 160 annually per *mu* of their land converted together with food rations. Although this kind of compensation appears to be more appealing to the peasants than farming itself, it is only applicable for eight years; after that, their livelihoods would again become problematic. This is exacerbated by the fact that it is not uncommon in this county as well as in many parts of China that many households do not receive their subsidies in cash in time and in full – an indicator of poor management of the funds earmarked for this specific programme by the local government (Bi, 2010).

In any case, this tiny amount of subsidy was not fully justifiable as a means of alternative livelihoods for the peasants when they are facing rising costs of living and production materials. Moreover, given the climatic conditions in this

---

5   According to the Grassland Monitoring and Supervision Centre of the Ministry of Agriculture, this circular sends a clear message to corrupt officials involved that the Grassland Law must be upheld. For the details of the cases, see http://www.grassland.gov.cn/grasslandweb/Article/ShowArticle.asp?ArticleID=103

dry region, it takes approximately 15 years for the trees to grow. Thus, how to take care of the trees in the long-run has not been well thought of. As a result, some planted trees even died off a few years later. And some forest land was even re-converted to farmland by the peasant households whose concerns about losing land is posing a threat to their livelihoods as their family sizes increase. As reflected by the county Forestry Bureau, peasants' lack of ownership over the trees put the sustainability of the forest programme at risk. In the end, balancing the interests of the local government and peasants sector poses a severe challenge to the county decision-makers (see Strauss, 2009).

To address peasants' disincentives in the forest land owned by the village collective, as another example, the Forestry Bureau follows the pilots of the collective forestry tenure reform in southern China, where cases of success are documented.[6] In fact, the effects of the pilots adopted in the South were taken up by the State Council, which led to the promulgation of the 2008 Opinions on Comprehensive Collective Forest Land Tenure Reform. Essentially, this policy is to emulate the HRS in the management of collective forestry land and peasants' ownership of forests across the country. It is seen as a major measure to boost the enthusiasm of the peasants, increase their incomes and make forest management sustainable. It stipulates that production and management of forests should be entrusted to individual peasant households by issuing extendable 70-year forest land use contracts, while the nature of collective ownership should be maintained. It calls for ensuring equal access to peasants' forest land rights and guaranteeing their rights to know and participate in the decision-making process affecting their land rights. Moreover, peasants are allowed to transfer, lease or mortgage the forest land use rights within the tenure period. Local government is asked to extend financial institutional services to the needy and establish forestry insurance to protect the peasants from natural disasters. This reform also makes an explicit call for strengthening public services to support forestry cooperatives and enterprises, which should play a leading role in forest management and production to promote economies of scale. It was set to be completed over a five-year period, during which forest land rights certificates should be issued to the individual households based on the registration of their contracted forest land (Xinhua News Agency, 2008).

In fact, long before the promulgation of this policy, the Forestry Bureau had tried to emulate similar measures to undertake collective forestry reform – with very limited success. The slogan of 'strengthening individual households' awareness and forest management ownership' was used to rationalize the policy and to motivate the peasants. However, the latter did not perceive this policy as something new, as the forest had already been under the management of the

---

6   Since 2003, collective forestry reform has been piloted in Fujian, Jiangxi, Liaoning and Zhejiang provinces. China has 2.55 billion *mu* of forest land (equivalent to 60 per cent of the country's total) under village collective ownership with more than half of the population living in these areas (Xinhua News Agency, 2008).

collective. Moreover, granting them long-term forest use rights might incur additional institutional burdens to shoulder, since they do not expect to gain benefits from the forest products which are not as marketable as in the past. As an official of the bureau indicated,

> The collective forestry reform here cannot be compared with that of the South, where the peasants can simultaneously plant other economic crops with the forest. Here, the climatic conditions simply disallow this. That is why it is not attractive to the peasants. So, we may not be able to continue the reform later on. (Interview with Forestry Bureau official, July 2008)

This remark proved to be realistic. While the use rights for a large proportion of the collective forests were said to be granted to individual households, the latter were reluctant to receive the use rights certificates. How to divide and distribute the collective forests to individual households is never an easy task for the Forestry Bureau. To a large extent, forest land distribution should be carried out on the basis of equitable conditions in terms of the quality of the forest land and the number of household members. In cases where poor quality land cannot be allocated easily, it should be done through lease, tender and auction to whoever needs it. When this cannot be realized, it rests with the village collective to dispense with. All these steps require sound planning and participation of the community in deciding on how the forest land can best be used and managed. Moreover, the peasants need to know whether the forests contracted to them are of economic value. The lessons from this county and other regions indicate that in most cases, local governments pay insufficient attention to the needs of the households and collective efforts (Miao and West, 2004). As a result, the reform can easily incur discontent and even conflicts among the various stakeholders involved, which deserves further study.

The lack of capacity of the Forestry Bureau to control inappropriate forest land use overshadows its overall reform agenda. Some staff blamed grazing as the number one threat to forest sustainability and emphasized the need to put a complete stop to it by severely penalizing those responsible, especially the households with a large number of livestock. Obviously, a lack of coordination between different line agencies constrains any conceived efforts in sustainable forest land management. These factors are compounded in the ongoing process of decentralization, in which local government lacks resources to implement the reform agenda (see Lieberthal and Lampton, 1992). What mechanisms should be established to empower the peasants to keep the local cadres in check remain unclear. As a result, 'the leadership of the rural collective, including the Party secretary, the village head and other village committee members, may co-operate and pursue personal interests as a collective' (Cai, 2003: 668).

In short, policies of grassland and forest protection and utilization have been mutually exclusive. The local peasants could only resort to short-term gains, sometimes at the cost of these resources (Cui and Wang, 2006). A lack of coordination in land use planning and management contributes to the failure of the programmes in which the peasants are caught in a vortex of uncertainty regarding the changes in land use imposed by the government. The relationships between natural resource tenure reform, poverty and the environment have not been sufficiently addressed in an integrated manner by the local government. The reform of land resource tenure from collective to household-based institutions signifies a simplistic approach that departs from the biophysical, economic and political constraints. As a result, even though new policy targets can be met on paper, the quality of natural resource regeneration and protection programmes can be put at risk.

Conclusion

This study indicates that the conventional individualistic approach to natural resource tenure has not brought about viable solutions to addressing the complexity of rural poverty and its underlying institutional constraints. Relying on market-oriented development models, China's economic success has incurred severe social and environmental costs. To a certain extent, the replacement of the collective institution of the commune with the HRS has exacerbated its developmental dilemma. The serious flaws of this approach and the rhetoric of development policies lie in their unresponsiveness to local biophysical, political and economic realities (Gupta, 1998). The HRS results in short-term development gains rather than sustainable resource use in the long run. Moreover, it is interwoven with the absence of appropriate institutional mechanisms for effective sustainable resource use and management in a region where poverty and natural resource degradation prevail.

The HRS has been a contributing factor for the fragmentation of social relations and the lack of capacity of the local state and lack of collective action among local communities in the development processes. The local state puts overt emphasis on meeting higher-level state demands rather than serving the community needs first (see Christiansen and Zhang, 1998; Kung et al., 2009). This is reinforced by the HRS whereby communities are marginalized in land use planning and the broader level of village governance. In the context of rising social inequality across the country, it becomes more difficult for the state to organize the peasantry. The fact that some peasants support the HRS actually implies that they mistrust other possible measures imposed upon them by the state. In other words, they are not given the space to explore other systems of land tenure by the state.

The absence of mechanisms for genuine peasant participation in policy-making processes concerning their land use and governance has yet to be addressed. It is important to understand how mechanisms of power have been able to function within society and between society and state in order to investigate the agents responsible for social constructs (Foucault, 1986). This chapter shows that although mechanisms for peasants' collective action remain unclear and even weak, their daily struggles are omnipresent in their disorganized and silent contestations. Their nascent resistance could lead to more organized institutions with clear motives and goals (see Scott, 1985).

As the peasant–local state relations become more murky and complex, and social dynamics among different social groups become more intractable, how to revitalize the Chinese countryside by making the complex social, political and economic relationships work for the goal of sustainable natural resource use and rural development remains an ultimate challenge for policy-makers. It is essential to tackle the structural barriers to law and policy-making mechanisms that put constraints on the representation of the peasants, whose need to participate in the process should be nurtured and empowered (Cai, 2003; Li, et al., 2004).

No single type of landownership is a blueprint for sustainable natural resource use and management (Dietz, et al., 2003). A new form of land tenure congruent with local economic, ecological, political and social conditions should be explored and tested by policy-makers. This form of tenure has to serve the imperative of sustainable resource use and management, which needs to be imposed on state–peasant relations. Thus, future natural resource policy developments should be based on sound analyses of these interconnections, the implications of which are vital to accomplishing sustainable development goals in China's dry north in particular.

References

Banks, Tony, Richard, Camille, Li, Ping and Yan, Zhaoli. 2003. "Community-based grassland management in western China: Rationale, pilot experience, and policy implications", *Mountain Research and Development*, 23 (2), 132–40.

Bi, Lei. 2010. "北京防沙最后的屏障: 河北沽源县生态移民调查" (*The last protective screen for Beijing's prevention of sand storms: A survey on ecological migration in Guyuan County, Hebei Province*), http://www.tieba.daidu.com/f?kz=951350767, accessed 13 August 2008.

Bramall, Chris. 2004. "Chinese land reform in long-term perspective and in the wider East Asian context", *Journal of Agrarian Change*, 4 (1, 2), 107–41.

Cai, Yongshun. 2003. "Collective ownership or cadres' ownership? The non-agricultural use of farmland in China", *The China Quarterly*, 175 (175), 662–80.

Chang, Kyung-Sup. 1994. "The social-political processes of agricultural decollectivisation in a North China commune", *China Report*, 30 (4), 395–406.

Chin, Gregory T. 2005. "Securing a rural land market: Political-economic determinants of institutional change in China's agricultural sector", *Asian Perspective*, 29 (4), 209–44.

Christiansen, Flemming and Zhang, Junzuo (eds). 1998. *Village INC.: Chinese Rural Society in the 1990s*, Honolulu: University of Hawai'i Press.

Cui, Haixing and Wang, Liqun. 2006. "退耕还林工程与农村社会运行: 基于河北省沽源县的实证分析" (Green for grain programme and rural society: An empirical study on Guyuan county), http://www.sociology.cass.cn/shxw/xcyj/p020061007006712652962.pdf, accessed 13 August 2008.

Dietz, Thomas, Ostrom, Elinor and Stern, Paul C. 2003. "The struggle to govern the commons", *Science*, 302, 1907–12.

Foucault, Michael. 1986. "Disciplinary power and subjection", in Steven Lukes (ed.), *Power*, New York: New York University Press.

Government of China. 2002. *Grassland Law of the People's Republic of China*, Beijing: Ministry of Agriculture.

Gupta, Akhil. 1998. *Postcolonial Developments: Agriculture in the Making of the Modern India*, Durham: Duke University Press.

Guyuan County Government. 2003. 沽源县志 (Guyuan County history), Baoding: Zhongguo Sanxia Press.

Guyuan Poverty Alleviation Office. 2007. 沽源扶贫开发工作情况的调研报告 (Research report on poverty and development in Guyuan County), unpublished report.

Hardin, Garrett. 1968. "Tragedy of the commons", *Science*, 162, 1243.

Hinton, William. 1990. *The Privatization of China: The Great Reversal*, London: Earthscan Publications.

Ho, Peter. 2005. *Developmental Dilemmas: Land Reform and Institutional Change in China*, London and New York: Routledge.

Hu, Wei. 1997. "Household land tenure reform in China: Its impact on farming land use and agro-environment", *Land Use Policy*, 14 (3), 175–86.

Huang, Yiping. 1998. *Agricultural Reform in China: Getting Institutions Right*, Cambridge: Cambridge University Press.

Jiang, Hong. 2006. "Decentralization, ecological construction, and the environment in post-reform China: Case study from Uxin Banner, Inner Mongolia", *World Development*, 34 (11), 1907–21.

Kahrl, Fredrich, Roland-Holst, David and Zilberman, David. 2005. *New horizons for rural reform in China: Resources, property rights and consumerism*, Stanford Centre for International Development, http://are.berkeley.edu/~dwrh/Slides/SCID_KRHZ_100105.pdf, accessed 16 August 2008.

Kung, James, Cai, Yongshun and Sun, Xiulin. 2009. "Rural cadres and governance in China: Incentive, institution and accountability", *The China Journal*, 62, 61–77.

Lai, Lawrence Wai Chuang. 1995. "Land use rights reform and the real estate market in China: A synoptic account of theoretical issues and the property rights system", *Property Management*, 13 (4), 21–8.

Li, Xiaoyun, Zuo, Ting and Ye, Jingzhong (eds). 2004. *2003–2004 Status of Rural China*, Beijing: Social Sciences Academic Press (China).

Lichtenberg, Erik and Ding, Chengrui. 2008. "Assessing farmland protection policy in China", *Land Use Policy*, 25, 59–68.

Lieberthal, Kenneth and Lampton, David. 1992. *Bureaucracy, Politics, and Decision-Making in Post-Mao China*, Berkeley and Los Angeles: University of California Press.

Lin, George and Ho, Samuel. 2003. "China's land resources and land-use change: Insights from the 1996 land survey", *Land Use Policy*, 20 (2), 87–103.

Miao, Guangping and West, R.A. 2004. "China's collective forestlands: Contributions and constraints", *International Forestry Review*, 6 (3–4), 282–98.

Ministry of Agriculture. 2006. "农业部办公厅关于河北省沽源县开垦草原案件查处情况的通报" (Circular of the Ministry of Agriculture's decision on the inspection and punishment concerning the Guyuan County of Hebei Province's mismanagement of grassland cases), http://www.chinaacc.com/new/63/73/142/2006/9/wa4652021614960023920-0.htm, accessed 16 August 2008.

Muldavin, Joshua. 1997. "Environmental degradation in Heilongjiang: Policy reform and agrarian dynamics in China's new hybrid economy", *Annals of the Association of American Geographers*, 87 (4), 579–613.

Nonini, Donald. 2008. "Is China becoming neoliberal?", *Critique of Anthropology*, 28 (2), 145–76.

Sanders, Richard. 1999. "The political economy of Chinese environmental protection: Lessons of the Mao and Deng years", *Third World Quarterly*, 20 (6), 1201–14.

Scott, James. 1985. *Weapons of the Weak: Everyday Forms of Peasant Resistance*, New Haven: Yale University Press.

Solinger, Dorothy. 2002. "The floating population in the cities: Markets, migration, and the prospects for citizenship", in Susan Blum and Lionel Jensen (eds), *China Off Centre: Mapping the Middle Margins of the Middle Kingdom*, Manoa: University of Hawai'i Press, 273–90.

Strauss, Julia. 2009. "Forestry reform and the transformation of state capacity in fin-de-siècle China, *Journal of Asian Studies*, 68 (4), 1163–88.

Szirmai, Adam. 2005. *The Dynamics of Social-Economic Development: An Introduction*, Cambridge: Cambridge University Press.

Wang, Shiying, Cai, Qiangguo and Gao, Yanchun. 2005. "Recent land use changes in north dust storm source area affecting Beijing and Tianjin: Agro-pastoral transitional zone", *IEEE International Geosciences and Remote Sensing Symposium Proceedings*, 2402–5.

Williams, Dee. 1996. "The barbed walls of China: A contemporary grassland drama", *Journal of Asian Studies*, 55 (3), 665–91.

Xinhua Net. 2006. "沽源: 非法开垦草原事件的背后" (Guyuan: What is behind the case of illegal grassland exploitation), http://www.he.xinhuanet.com/dishi/2006-03/30/content_6607521.htm, accessed 20 August 2008.

Xinhua News Agency. 2008. *China to push forward collective forestry land reform*, http://www.china.org.cn/government/central_government/2008-07/15/content_16009116.htm, accessed 20 August 2008.

Yeh, Emily. 2004. "Property relations in Tibet since decollectivisation and the question of fuzziness", *Conservation and Society*, 2 (1), 107–31.

Zhang, Li. 2001. *Strangers in the City: Reconfigurations of Space, Power, and Social Networks within China's Floating Population*, Stanford: Stanford University Press.

Zhou, Jinhai. 2010. "浅论沽源县农民专业合作社发展" (Brief account of specialized farmers' cooperatives in Guyuan County), http://www.cfc.agri.gov.cn/cfc/html/141/2010/20101202150621171402136/201012021506211714021 36_.html, accessed 25 August 2008.

Chapter 6

Land Transfers for Scaled Development: An Economic Fix or Marginalization of the Poor?

Introduction

The preceding chapters provided a snapshot of the evolution of China's land tenure reform especially concerning the remarkable changes marked by the introduction of the HRS in the wake of the market reform. Acknowledging the indispensable role of the HRS in facilitating growth and social and political transformation, it highlights its induced institutional challenges for sustainable rural development. Against this backdrop, this chapter continues by providing a more detailed introduction to the current land tenure reform policy trajectories and local experimentations in using the HRS as a basis, as well as the associated major scholarly debates to further shed light on the linkages between land tenure, development and village governance for possible institutional readjustments for more pro-poor development outcomes.

As mentioned earlier, the mainstream land policy direction and scholarly thinking treats land as an essential element for fast growth or economies of scale, and how to create a land market favourable for land transfers is the key. In essence, this implies that more and more peasants should vacate their land for those who are more capable of farming, thereby contributing to more efficient land use and the flow of rural population to the cities. And this is seen as fundamental to continued economic growth and balanced regional development, as these peasant populations or migrants in the cities have the potential to significantly boost the domestic consumption desperately needed for China's economic transformation towards more humane, equitable and sustainable goals.

In essence, this discourse is centred on the magic of the market in fixing the current malfunctioned land management system especially concerning farmland expropriation whereby land loss and inadequate compensation for lost livelihoods have become paramount. As such, it is widely proclaimed that more clarified collective landownership and land use rights should be given to the peasants and a better functioning land market should be created to allow for more direct engagement by peasants with the market and thus avoid the land market being handicapped by other actors, especially the local state. Obviously, the general land policy trajectory rests with the development need for quick growth and poverty alleviation outcomes through emancipation of the peasants from their land. This is further supported by local government land institutional innovations in an attempt to provide more attractive compensation packages to the peasants whose willingness to vacate their land is of vital importance for the local state and business enterprises to strike a quick deal.

Underlying this approach is the so-called urban–rural integration agenda which received a major boost by the 2010 No. 1 Document on more balance and integrated urban–rural development as a way to increase peasant incomes and to enhance scaled production. This policy paper calls for a more coordinated regional and national development, which is based on industrialization, urbanization and agricultural modernization.[1] In following suit, local governments have enhanced their efforts in promoting land shareholding cooperatives introduced even as early as the 1990s in southern China. The notion of land shareholding cooperatives refers to the institutional arrangements for individual peasant households to transfer their land as a share to the direct management of cooperative organizations or agribusinesses. The peasants can either remain in their land and may be employed by the latter, or simply move out of land for other opportunities. In both cases, in principle, they should be able to collect dividends as shareholders. It is seen as a land institutional innovation by the state in the use of market mechanisms to equalize urban–rural development economic structural disparities in systematic rural transformation.

Being implemented throughout the country, the urban–rural integration agenda involves different forms of land transfers such as land swap, lease, social insurance for land, and land shareholding cooperatives. It is not possible to provide a comprehensive description and analysis of all these institutional arrangements, but given their commonality in the role of the local state and advantaged groups in organizing peasant–land relations, this chapter pays particular attention to the development of land shareholding cooperatives especially in respect of different discourses, the underlying contradictions revealed by local practices, and the implications for urban--rural integration in general and land tenure reform in particular. Simply there has been a lack of study of why the local state is in favour of the market-oriented land tenure reform approach as they are often blamed for the problems created in relation to land tenure insecurity. In other words, if the local state has played a double role of institutional innovator and trouble maker, to what extent can the ongoing local land shareholding experimentations in line with market principles and approaches be successful? This chapter provides a critical lens for the study of land institutional change by investigating the key challenges for peasants' self-initiated reform patterns, which are crucial for both the market economy and peasants' sustained livelihoods that largely depend on land governance. It concludes that China's land issues are not solely contingent upon who owns what, but more about who decides on what. What matters more is governance, which ought to give full play to peasant participation despite the

---

1   More information on this document can be found at http://www.chinanews.com/cj/cj-plgd/news/2010/01-31/2101018.shtml

role of the market in regulating the state and other actors. Peasant participation in sustainable land use and management can only be meaningful if it is coupled with drastic enhancement of sound village democratization mechanisms. Without empowering the poor in land shareholding arrangements, farmland loss and poverty can even be exacerbated by the cooperatives.

China's Land Reform at a Crossroads

China's land reform has reached a critical stage where drastic institutional changes are needed to incorporate local communities into land use and management regimes. Existing institutions have provided ineffective mechanisms for peasant-centred decision-making processes, which are essential to the identification of appropriate forms of land tenure arrangements. For many scholars and policy-makers, the HRS seems a stumbling block to the realization of economies of scale due to its nature of farmland fragmentation. As smallholders lack adequate access to inputs, technology, information and markets, their produce can hardly meet the demands of final consumers (Hu et al., 2007).

The central government has conceded considerable flexibility to the local governments to test innovative practices in rural and urban development. Deng Xiaoping's visit to Guangdong Province in 1992 significantly boosted bolder economic reforms in the region, which set an example for the rest of the country. His visit complemented the government's vision of modern specialized peasant collective organizations as the second step towards agricultural leapfrogging, with the HRS as the first step (Zhang and Donaldson, 2008). After Deng's visit, local governments felt the urgency to drastically accelerate economic development on a large scale in the pursuit of growth at all costs. As a result, large tracts of farmland were taken from the peasants for many purposes, among which included agricultural modernization (Zweig, 2000).

In parallel with the HRS, since the 1980s, country-wide grass-roots experimentation in land use and management is exemplary of the increasing role of land shareholding cooperatives, especially in the relatively developed eastern and southern regions. This is led by the entrenched market reform in the countryside, where land value has been on the rise. In this system, peasant households are encouraged to transfer their land use rights to other parties such as agribusinesses via the village administrative committee (which is in charge of improving land use and in particular deriving benefits from the land). As such, to a large extent, the peasant households become land shareholders in the village administrative committee's businesses which are managed on behalf of the shareholders.

Furthermore, the current urbanization rate of 47 per cent has thus become a catalyst for a new wave of 'land enclosure or town creation movement' to unleash the huge consumption potential of the vast rural migrants and residents in the cities and thus to boost national economy. This is deemed as the key to the economic transformation agenda of the 12th Five-Year Plan for National Development (2011–2015) in general and the urban–rural integration agenda in particular (see Chi, 2010). As such, ongoing local experimentations have opted for

> ... "swapping land for social insurance" schemes to enable peasant households to relocate to the cities and thus become urban residents living in high-rise buildings, which is a symbol of better livelihoods and modernization. However, this has been a daunting challenge for China given the rising difficulties for the cities to create enough employment opportunities, affordable housing and medical and social care schemes to the new urban residents and rural migrants who seek temporary livelihoods. Moreover, land transfer from the hands of individual peasant households to other state and business parties lacks firm legal backing and procedures, for in many such cases peasants are not given the rights to negotiate the terms and conditions offered. And those seemingly favourable social insurance schemes are not equivalent to the value of land property. (China Institute for Reform and Development, 2010)

China's land law and policy developments have been developed to facilitate land commercialization as the market value of land drastically increased in the reform era. The current land tenure system actually serves this purpose. Land remains the property of the state and rural collective, and its sale is forbidden. Only under state acquisition can the farmland be converted into non-agricultural purposes.[2] Any plans for such a conversion must undergo a government approval process. But this stipulation excludes the cases where rural collective non-farmland or construction land can be legally acquired for the development of village enterprises, housing and infrastructure. Zhejiang Province is a pilot region to carry out the promotion of rural collective construction land transfer. This move is seen as a major step towards equalization of rural and urban land rights, which means that rural land should be allowed to be transferred in the same way as urban land. It is expected that this act would provide favourable conditions for those who can invest in this sector to bring about improved land use efficiency.

As a result, local governments and village collectives are offered ample space for direct manipulation of land use. Although there is no formal rural land market, in regions where there is a high demand for land, the village collectives developed various tactics to transfer or lease the collective land to outsiders for non-agricultural use in the informal market. To a certain extent, a land shareholding cooperative is the invention of the alliance between business enterprises, local governments and village collectives to sidestep the existing legal and policy barriers to the formation of a formal land market.

It is through the shareholding cooperatives that the village collectives and local state can relatively easily lease land to businesses. The latter can bypass the

---

2   In urban areas, land use rights can be traded on the market, but landownership rests with the state.

individual peasant households and directly engage the village collectives and local government. In the name of granting the shareholders more equitable economic benefits accrued from such arrangements, village collectives can use this system as a better excuse to acquire land. In this way, an informal market is created, whereby land prices and economic compensation paid to the affected households are dictated with inadequate peasant participation and discretion. In fact, many local governments have legalized it to facilitate urbanization and industrialization. For instance, in Dongguan, Guangdong Province, in 2004 the local government drew up a strategic plan to complete rural shareholding reform in three years and transform all peasants into urbanites within five years. In Kunshan, Jiangsu Province, by 2004, 142 cooperatives accounting for one-tenth of the province's rural population were established with each household receiving at least 10 per cent of the value of their shares (Po, 2008: 1615).

Land shareholding cooperatives has been promoted as an innovative fix to radically achieve optimal land use, scaled agricultural production, rural development and collective decision-making. Although individual peasant shareholders can opt to continue farming by themselves, in most cases they either leave the village for the cities or seek all possible sorts of work in factories. Despite the fact that these arrangements have yielded some positive impacts on land use and peasant livelihoods, one cannot overestimate their roles in promoting rapid agricultural development without taking account of the complex linkages between land and village governance. The reason for a lack of built-in mechanisms for community participation in the Chinese countryside clearly resonates with the slow progress made in village democratic governance, as a precondition for properly functioning cooperatives. These cooperatives are interwoven with the complex relationships between the people and the state, and the particularities of a centralized system of governance (Plummer, 2004). State-dominated collectives actually obstruct peasant power and choice over the management of the cooperatives and village governance. Moreover, the embedded institutions, that is, informal rules of the game, customs, traditions, norms and even religion, further complicate the formal institutions such as bureaucracy, policy and the judiciary (Williamson, 2000). As a result, China's land reform has perpetuated a form of state domination that puts the vulnerable poor on the margins of policy, law and the rural development reality. Although land shareholding cooperatives have received a major policy boost as the market reform sets in further, they are far from effective and hardly resemble pro-poor institutions for sustainable rural development.

In retrospect, the differentiation between land shareholding cooperatives and the People's communes enacted in the 1960s are remarkable. The latter was even a higher-level cooperative organization in terms of scaled operation across villages and townships. The government managed to transform the majority of traditional marketing and supply organizations into the so-called modern cooperatives. However, these hastily-established institutions failed because of gross inadequacies in the facilities, resources, skills and experience of the government, among many other factors (Skinner, 1965). The state and the commune played a dominant role, whereas individual households joined the commune as members, not shareholders. But in this sense, landownership and use rights were clearer than in the current case. Although the current land shareholding system is based on the HRS which allows individual households to own land shares, it may even complicate the ambiguity of collective ownership of land. The roles of household shareholders in the governance of the cooperatives can be more difficult to define than in the case of the commune. As a result, a simplistic approach to land commercialization in the form land shareholding cooperatives may not necessarily constitute a viable solution subject to varied local contexts.

The central government has taken a cautious approach to land reform experimentations. On the one hand, it tries to stimulate market-induced mechanisms; on the other hand, it has to strengthen its grip on fighting corruption and misuse of power by the local government. As an official from the Chinese Ministry of Land Resources remarked on latest policy developments:

> The public media has spread the wrong message about the latest policy developments. In fact, no matter how peasants are allowed to transfer their land rights, this must be done within the limit of local and regional land use planning under the direct control of the village collective. This means that the peasants' rights to do so will still be determined by the village administrative committee and local government.[3]

Local Practices in Land Transfers

Of all the major cases of local experimentations on land shareholding cooperatives, Zhejiang Province was chosen by the Ministry of Agriculture to pilot peasant specialized cooperative organizations in the 1990s. By 2004, the number of these organizations reached 554,000 and the total number of peasant households involved reached 2,029,500. By 2008, 42 per cent of the households contracted land or 25.5 per cent of the total arable land had been transferred to those more economically and physically capable and sizeable households or business enterprises (Hu et al., 2007: 444). In response to the government's call for specialized zoning development for agriculture, this province has put ample emphasis on developing priority enterprises through fostering 'specialized households, specialized villages, specialized townships and specialized markets'. Wenling City has pioneered these initiatives and become the so-called 'hometown of oranges, sugar cane, water melons and grapes'. Grown in ecological demonstration sites, these cash crops are perceived to be crucial to scaled agricultural development, as more and more specialized large households can acquire land with the support of the government for cash crop farming. Simply put, land transfer greatly encouraged by the local

---

3   Interview with the Deputy Chief of the Division of Cadastral Management of the China Land Survey and Planning Institute on 13 June 2009.

government takes varying forms such as land shareholding cooperatives, sub-contracting, leasing, swapping and so forth. Service centres were also established to facilitate land transfer processes. However, to the provincial government, this change is simply not fast enough, due to four major institutional constraints: inadequate role of government support; lack of service support for more regularized land transfer procedures; lack of social insurance schemes for peasant households who are prone to farmland expropriation; and peasants' ideological backwardness on the importance of modern institutional changes (Zhu and Chen, 2008).

Furthermore, promoting the industrial mode of development in scaled agricultural production by the 'dragon-head enterprises' (*longtou qiye*), local governments often assume that this institutional arrangement would efficiently pool together economic, technical and market resources and overcome the disadvantages of the HRS. In Guangdong Province, by 2008 almost 20 per cent of the farmland had been transferred through sub-contracting, mortgaging and reversed sub-contracting. These arrangements paved the way for the creation of the land shareholding system (Dongfang Zaobao, 2008). Similarly, in Jiangsu Province, by 2008 there were nearly 12,000 peasant specialized cooperatives of all kinds involving more than 5,000 households or 35 per cent of the rural population. In particular, the experimentation in land shareholding cooperatives in Yangzhou City has been widely acclaimed as a major achievement in stabilizing and improving the current land use rights structure, enhancing high efficiency in large-scale agricultural production, raising peasants' incomes and promoting the government's latest development agenda – building the new socialist countryside. It is hoped that the functioning of these organizations will speed up village democracy, since village politics will eventually be shaped by the three relatively independent institutions – village party committee, village administrative committee and collective economic organizations. Such a model is regarded as an innovation that contributes to social stability and boosts rural economic development (Li, 2008).

As mentioned earlier, in practicing different schemes of land transfers, it is commonplace that the households with a lack of skills, funds and access to the market are encouraged to transfer their land to more capable actors. To facilitate this process, some village administrative committees bundle the contracted farmland from the households and then unilaterally re-contract the land to other households or enterprises. Those who give up their contracted land receive an annual rent from the village administrative committee. This model is called 'reversed sub-contracting' (*fanzu zhuanbao*). Normally, this model is aimed at the formation of land shareholding cooperatives, where the households tender their land use rights for shares. Yet, this has never been a smooth process, since many peasants are concerned about losing their rights completely after joining the cooperatives. A lack of secure rights further deters their incentives in becoming shareholders given the fact that many households in China do not have certified farmland contracts. Moreover, there is a high-level uncertainty over the adequacy of compensation and social benefits for land transfers among the peasant households (Zhu and Chen, 2008).

The Nanhai Model

A prominent case of land shareholding cooperatives is the Nanhai model, exemplary of many cases of land development supported by the central government. Nanhai district is under the jurisdiction of Fuoshan Municipality, Guangdong Province. Rural economic reform was started in 1987, when the State Council designated Nanhai as one of the major sites for demonstration pilots in large-scale agricultural production whereby land transfer was permitted. At that time, the HRS had gradually become a major constraint to organized farming. Coupled with structural changes in agriculture where farming was no longer appealing to the peasants, rural–urban migration was phenomenal. More than two-thirds of the peasants sought employment in the cities and left their land either fallow or semi-unattended. Agricultural production had slowed down substantially and could not be boosted because of peasants' low incentives and a shortage of labour. Rapid urbanization also induced a high demand for farmland. In turn, land expropriation to facilitate urbanization caused mounting peasant discontent over inadequate compensation and its negative impact on their livelihoods (Wang and Xu, 1996).

   To the Nanhai government, the reform was nothing more than agricultural modernization through land concentration, or outright land amalgamation. At an initial stage, peasants' entitlements to their contracted land were adjusted. Those peasants, to a large extent the out-migrants, who did not rely on land for their livelihoods were not allowed to keep the land. Even those who possessed land use rights certificates were asked to give up their land for tender arrangements through which land use rights changed, as large-size households with more capital and skills in agricultural production, or those small businesses with vested interests in the land, received the land. In essence, all the peasant households were encouraged to transfer or give up their land to others, backed by the local government's promise to provide social security such as subsistence and employment conditions. And large-scale land transfers were encouraged to make land available for scaled production. Village administrative committees acted as managers of the cooperatives in converting the collective land and assets into land shares for non-agricultural purposes such as building factories as well (Wang and Xu, 1996; Jiang and Liu, 2004).

   Although this storm of land reform measures created favourable conditions for large-scale agricultural production, the local government felt that it had not met their expectations. They believed that the speed of the land amalgamation process remained slow, and had not fundamentally changed the situation when small-scale and fragmented farming existed. This was demonstrated by the large number of peasants who wanted to remain in the villages for farming. With rapid rural population growth, the average size of households declined substantially. At the same time, peasants raised their awareness of the increased value of land.

Although they did not fully rely on their land to meet their subsistence needs in contrast to those living in poorer regions, they were still inclined to hold on to their land for future use (Wang and Xu, 1996).

Because of these difficulties in furthering the reform, the local government started testing a peasant-centred institution to cope with their unwillingness to vacate the land. With disbelief in land privatization and nationalization, the reformers decided to establish the so-called socialist cooperative economic institution, that is, peasant land shareholding cooperatives for comprehensive rural economic development outcomes. In 1992, a demonstration pilot was started to enable the peasants to become land shareholders by granting them economic and operational rights in different shareholding arrangements. Then the local government managed to consolidate and accumulate the fragmented land and put it to different uses such as the establishment of zones for farmland protection, industrial development and public utilities. By 1995, a total of 1,869 rural shareholding organizations had been established, 80 per cent of which were at the village level (Zhao, 2007: 40). It is important to note that the local state played a crucial role in initiating and managing the cooperatives. In most cases, these cooperatives were transformed into agricultural development shareholding corporations under the auspices of the local state. The latter, through its policy and funding support for the cooperatives, provided ample space for their development. As rural development required collective action to a certain extent, the local state–peasants–business alliance was important in overcoming many difficulties in the farming and marketing of agricultural produce (Wang and Xu, 1996). In addition, being the holders of 50 per cent of the cooperative shares, village collectives play an essential role in channelling the funds from higher-level government for public service delivery and social welfare provision to their fellow households (Jiang and Liu, 2004).

Moreover, to enable the peasants to join the cooperatives, each peasant in Nanhai received an equal number of land shares. In some cases, other types of shares based on labour contribution and land use period were also introduced to further stimulate peasant incentives. Three different models were developed to cater for local circumstances in three townships. For instance, in the Lihai model, peasants' shareholding rights were hinged on the land and other assets contributed, which could be transferred, mortgaged, inherited and even given out as gifts within the organization. In contrast, the Guicheng and Pingzhou models did not allow for trading of land shares except for inheritance purposes. But in all these models, the distribution of dividends was decided by the shareholders' assembly, which was held regularly with the participation of the majority of the shareholders. All management information especially those related to financial management of the enterprises was supposed to be released to the shareholders who had the exclusive rights to judge whether there were issues related to lack of transparency, fairness and operational efficiency. Around 40 per cent of the annual revenues were distributed to shareholders as dividends, while the other 60 per cent was kept in the cooperative for public use and enterprise redevelopment (Huang, 2005; Zhao, 2007).

Despite the proclamations of success by the local government, till now, the developmental effects of the Nanhai model are not as far-reaching as expected. Due to the continuous shortage of natural resources such as water and farmland, how to maximize the utilization of these resources to suit the need of development has posed a challenge for the local government. For instance, more profitable sectors such as vegetable- and flower-growing businesses are targeted with a view to developing a modern urban agricultural sector. However, as these agribusinesses have a huge demand for farmland, food security is put at risk. The local government seems aware of this fact and even calls for maintaining the current level of food supply, but has no other considerations than encouraging the dragon-head businesses, normally part of the land shareholding cooperatives, to play a leading role in agricultural modernization as a solution.

Agricultural development has been slow. This is further constrained by the weakness of the peasants in organizing more effective land shareholding cooperatives as originally assumed (Nanhai Agricultural Bureau, 2007). According to the 2003 statistics, the average per capita revenue of shareholders was only RMB 1,180. Actual revenue distribution was also skewed, contrary to the charters of the cooperatives. In south-eastern areas of the municipality where industrial enterprises were paramount, the average revenue per capita was almost ten times higher at RMB 15,000 than other locations. For instance, in western areas where agricultural enterprises were the majority, the peasant shareholders had not received any of the profits distributed (Huang, 2005).

Local officials and peasants have perplexing attitudes towards the model. Some officials do not see a marked difference between the land shareholding cooperative system and the HRS, which does not pose a problem because it serves their interests as long as they have the final say over the land. By contrast, many peasants complained about the frequency of government policy changes. When the HRS was implemented, the local government talked about its merits in safeguarding the long-term rights and interests of the poor. In a similar vein, they confused the peasants by saying how profitable the land shareholding system would be. But the peasants showed their discontent over the loss of their land to the cooperatives, whilst under the HRS at least they had some land to meet their subsistence needs. Once the land had gone, they would end up losing the safety net, for their benefits from the cooperatives were uncertain. They simply doubted that the cooperatives would be managed well by the village leaders and agribusinesses. They were unsure about whether the land should be distributed to individual peasants or re-collectivized in the hands of the so-called peasant–business–local state alliance, but they had no choice but to listen to what they were told from the top (Wang and Xu, 1996).

The inefficiency of the Nanhai model has been criticized for its inflexibility in allowing the peasants to trade their land rights freely in a market that should be created. As their land shares are administered at the village level, it is questionable whether the village administrative committee can be economically efficient in maximizing profits for their constituents. By contrast, other regions of China have tried to adapt the Nanhai model to deepen market-oriented mechanisms, among which is the experimentation on complete separation of peasants' land use rights from the village collective landownership. This means that they try to avoid the involvement of the village administrative committee by putting more emphasis on empowering the shareholders in more direct land share management, as the village committee is often found to be corrupt in land governance. For instance, in Wanfeng Village, Baoshan District of Shenzhen Municipality, soon after the market reform was started, the local government realized that the HRS had its inherent shortcomings, one of which was the difficulty to organize agricultural production and peasants' participation in local administrative affairs. Farming efficiency was low, which was a major incubator of chronic poverty. Constrained by a lack of revenue from agriculture, the local government could not deliver the social service as it was supposed to. All these factors prompted the local government to come up with a new slogan of 'building a socialist industrial village through common property ownership'. In essence, this institution was about shareholding arrangements based on the land and production materials of the peasants. But it provided more mechanisms for building equity in terms of complete coverage of peasants who were free to choose their preferred means of investments in the shareholding corporation. Even for those who were unable to contribute in cash, the corporation covered their shares and allowed them to enjoy the benefits as shareholders. As a result, the shareholders had equal rights and opportunities concerning the development of the corporation, since they were considered to be the more genuine property owners of the cooperative corporation (Huang, 2005).

Complex Power Relations and Ineffective Developmental Outcomes

According to Guo Shutian, former Director-General of the Policy, Institutional and Legal Reform Department of the Chinese Ministry of Agriculture, the land contractual use and operational rights are no longer vague, but landownership remains an empty shell. Opposing divergent views on land ownership – privatization, nationalization or perpetual land use rights, Guo holds the view that the land shareholding cooperative system can be innovative in filling this gap to ensure that peasant shareholders can decide on land use just as real landowners. As a result, they will be able to more effectively exercise their economic and political rights concerning their land. Thus, the land shareholding system is seen as potentially efficient in safeguarding their rights and preventing their land from expropriation (Li, 2008).

Land shareholding institutions play an important role in reorganizing the Chinese peasantry in agricultural production and village industrial development by taking full advantage of the market and business opportunities for peasant investments in land. However, when this occurs, pure land farming gives way to agribusinesses and non-farming sectoral development, which involves high stakes for the shareholders to manage benefits and losses. The effects on poverty alleviation and rural sustainable development are far from clear due to a lack of in-depth studies. Above all, these institutions, albeit successful to a certain extent, have not yielded hugely proclaimed benefits to the peasants.

Given the varying degrees of success and failure of the experimentation on land shareholding cooperatives in China, institutional constraints to more peasant-centred economic organizations need to be understood. In many cases, the organization of land shareholding cooperatives is based on informal institutions or peasant relations in the process of their development and operation. These rural social relations feature a mixture of some kinship and market rules, which make governance of these organizations complicated. In many cases, membership is closed in the cooperatives despite the stipulation of the charter on the policy of 'free entry and free exit'. Membership is carefully vetted by the leaders to ensure that the members meet physical and technical requirements. Quite often the members are not allowed to withdraw when the cooperative is experiencing profit or property losses. Share purchase is extended to each member with the cooperative leaders acting as their patrons. The numbers of shares one can purchase also depends on various factors, which in turn determine the shareholders' power in decision-making. In the case of transfer of property rights, it can only be done within the cooperative, not outside it. And it is the board of directors that have the power to decide on such transfers. Thus, the ownership title is a hybrid form between individual and collective titles for Chinese-style cooperatives (Hu et al., 2007: 449).

It is important to note that the heterogeneity of the cooperative membership is complex, since those with more shares have more decision-making power than the others. This means that the majority poor shareholders cannot exert much influence over governance. They have to give up their decision rights to the cooperative leaders–board of directors, most of whom are village leaders and business enterprises representatives. For instance, the general director can be a large shareholder, whose shares count for as much as 20 per cent of all shares in some cases. As a result, the normal shareholders cannot decide on the major issues of cooperative governance even if they attend the assembly (Hu et al., 2007; Zhang and Donaldson, 2008).

Land shareholding cooperatives involve the transfer of land rights among households and between households and other shareholders such as enterprises and local states. Quite often, these transfers are brought about with the infringement of peasants' rights, lack of legal stipulation on the protection of their rights, waste of resources and weak oversight of cooperative management. The local state and village leaders play a major role in accountable and transparent management practices, but may be unable to abstain from the use of power for private gains (Hoff and Stiglitz, 2004). According to Zhang (2009), the institution of the land shareholding cooperatives has little to do with the so-called private and collective landownership. Rather, it is all about transferring peasants' land to large land shareholders. Thus, for the peasants, landlessness may become a reality. It would be unrealistic to think that the peasant shareholders can fully enjoy their rights as shareholders, for their limited land shares mean minimal real rights in the cooperatives. Or simply put, the cooperative is an effective institution that exploits peasants' land and their rights. Given the evidence of land loss and conflicts arising from the operation of these organizations, the government is trapped in its effort to transform the rural economy. This is inextricably linked with the process in which village democratization takes place in China. It is in this process that the rights of the peasants are being renegotiated and the struggles for power between the peasants and their leaders occur in silent and non-silent forms.

Debates on Individual and Collective Choice Over Land Rights

To improve the accountability and effectiveness of the ongoing local experimentation with land shareholding cooperatives, again, the debates return to the land rights issue. Qin (2006) strongly argues that landownership should be assigned to the peasants, which is crucial to democratic village governance. Disagreeing with others who are concerned about the possible effects of land concentration in the hands of a mighty few as is evident in many developing countries, he contends that dealing with the unconstrained autocratic power of the state and business alliance that severely undermines the rights of the peasants is of vital importance. In this respect, land privatization characterized by free land transactions does not necessarily cause landlessness of the poor and concomitant peasant unrests.

Furthermore, to Qin (2006), the absence of an effective control system over state behaviour and channels for the peasants in collective negotiation in matters concerning their rights and participation in legal and policy-making processes exercise a major constraint on their capacity to hold the cooperative leaders accountable. They should have the freedom to form organized groups, which are stronger than local state-dictated village collectives to represent their collective rights in their participation in land governance. When this freedom is restricted, the formation of land shareholding cooperatives may not necessarily lead to organized peasant action in land use and management.

It can be seen that the study of land rights should pay more attention to exploring the political, social and legal conditions under which peasants' land shareholders rights can be created. However, it does not mean that China's land reform should opt for absolute privatization in order to solve the problem, because the latter does not exist even in Western countries. In fact, citizens' land rights are surrounded by more restrictions than as compared to other rights because of the state's need for land utilization. That is why land always embodies the state interests in realizing its potential to cater for the public interests. The state has the final exclusive right to land acquisition. The ultimate question is who can represent the peasants' vested interests in land use and negotiate with the state and businesses in this process.

To Qin (2006), this question hinges on the degree to which peasants' land rights can be redefined and strengthened. Just like the cases of those 'unique' villages where the local communities did not choose the HRS when China's economic reform started and insisted on the commune system,[4] the peasants ought to be given the flexibility to choose an appropriate land rights structure. This actually complies with the operation of the market economy, which should provide avenues for people to make economic choices. They ought to be given the space to organize themselves around land management on a voluntary basis. Qin's rural survey conducted in Hunan Province in 1997 indicated that almost 50 per cent of the respondents expressed their preference for more strengthened private land rights based on equitable principles. Solving this issue is of crucial importance to address the issue of who owns China's land and who has the exclusive power to do so, which is deemed as a fundamental question pertaining to China's land reform (see Ho, 2005). Peasants' landownership should be recognized by law – be it individual or collective ownership – as long as it is based on their choice. In the case of individual landownership, peasants can still form land groups. And in the case of group or common ownership, individual members should be allowed to withdraw their membership. Given the weakness of small-scale household farming in China in terms of mitigating their vulnerability to the market, it is important for the collective to play an essential role in organizing the peasants. But this collective must be formed according to peasants' own choices. To reach this end, the law should provide ample space for peasant organizations (Qin, 2006).

Qin's viewpoints are deemed to be simplistic by others. From a historical perspective, peasants' ability to make institutional choices cannot be overestimated. It is argued that the collectivization period of the 1950s–60s was marked by strengthened organization of the peasantry. By comparison, agricultural decollectivization under the HRS since the late 1970s has led to the disintegration of peasant communities (Zhou, 1996). The HRS has brought about increased rural societal differentiation in terms of inequality in incomes and access to social services. It has actually dissolved the basis of the socialist superstructure and removed the socialist economic base. Community ownership and management is needed to replace it (Hinton, 1994).

To others, however, the HRS does not indicate the weakening power of the peasantry. Rather, it has unleashed the market mechanisms for peasants' autonomy and self-control, in contrast to the top-down relations in respect of strict control on labour mobility and the considerable power of local cadres in the collectivization era. And the cause of increasing inequality has much to do with the underdeveloped market which has not fully reached many remote poor areas. In any case, rural differentiation is a serious problem for the government, as it is dividing rural society into different groups such as the peasantry, proletarians, capitalists and government officials (Knight and Song, 1993; Rozelle, 1994; Bramall and Jones, 2000).

---

4   The unique case of commune villages is addressed in Chapter 7.

Complex rural social differentiation has strong bearings on the role of the government. With the introduction of the HRS, the central government's control over local governments and its influence in rural China has weakened in the process of decentralization. As a result, local governments have played a larger role in economic development, and faced mounting challenges of revenue generation to support their programmes (Bramall and Jones, 2000). They have gained substantial political power over the masses, which triggers incessant cases of corruption and conflicts. Confronting the control of land rights, which power is the strongest is ultimately contingent upon the extent to which it can influence the other power holders. Given the unique features of physical and social fragmentation in the Chinese countryside, it is hardly possible for the small landholders to exert control over their land. In this situation, stabilizing their land rights proves to be a daunting challenge, which has complicated the formation of land shareholding cooperatives to accommodate the heterogeneous interests of the shareholders.

Furthermore, where biophysical conditions set limits to large-scale farming, any attempts to consolidate the HRS in farmland management can only exacerbate the existing land fragmentation. More land-induced conflicts can occur because of multiple claims over the same land plot. Tackling this problem would make way for the exploration of new institutions. Thus, if the reform is inclined to land privatization, the ruling group can again easily infringe the peasants' land rights (Wen, 2004; Cao, 2005; Cheng, 2006). Moreover, the institution of collective landownership has social roots which cannot be simplistically understood from economic and legal angles. It has its inherent social contractual and cultural meanings. Because of the rules implicit in land relations, collective land rights reflect the nature of bundles of rights and thus complicated social relations. As these relations are seemingly ambiguous for outsiders, any attempt to clarify the current land rights structure will not lead to the stabilization and securing of the peasants' rights. Even though the law gives the peasants all the rights that they may deserve, it does not necessarily mean that these rights are safeguarded and enforced effectively. Land rights in any country have never been exclusive to the landowner, land cooperative, the state or other entities. Facing the challenges to food security, the state has become more important in intervening in land use to ensure the production of quality food and agricultural produce. In this sense, the market can never play the role that some deem as the best instrument in regulating land use. Given the weakness of the small landholders in agricultural production, it is important to organize the peasants in order to realize the full potential of the market, the peasants and the state in satisfying the national need for the preservation of farmland, thus ensuring national food security.

State institutions need to understand the local conditions and the law must be able to play an essential role in ensuring its effective enforcement to safeguard the rights of the affected. Moreover, to make the law work for the poor, democratic village governance (in respect of village elections for instance) ought to be improved to ensure effective legal changes. This also means that legal reform to strengthen property rights should be enhanced by mechanisms to hold leaders accountable (Deininger and Jin, 2009). It is no wonder that the current land shareholding cooperative systems present a last resort for the Chinese government to safeguard peasants' rights and maximize land use efficiency, despite many cases of failure due to a lack of mechanisms for assuring the effective execution of peasant shareholders' rights in cooperative organizations and village governance as a whole. It is far too early to ascertain their impact on village governance, a topic which requires further studies (Po, 2008).

In a nutshell, land shareholding cooperatives have largely been political in nature and ideologically earmarked as the last means of striking a balance between socialism and capitalism, or land nationalization and privatization. In other words, they embody the so-called unique feature of Chinese economy and society, that is, socialism with Chinese characteristics. On the one hand, they facilitate the establishment of land markets characterized by local state domination in land transfers. On the other hand, they pool the land resources of individual households together under cooperative operation and management, which exhibits the socialist feature of collective action. In this way, the roles and responsibilities of the state, the village collective, corporations and peasant shareholders can be ambiguous depending on specific local circumstances. It is this ambiguity that facilitates the development of the quasi-land market, which serves the state's interest to exert sufficient control over the cooperatives. Although relevant laws and policies seem to promote greater autonomy for these organizations, peasant shareholders' rights are not automatically improved due to the unbalanced power of different stakeholders. This means that the actual effects of these organizations on sustainable rural development are uncertain.

These debates have repercussions on the ongoing urban–rural integration agenda. It is argued that large-scale farming or mechanization does not necessarily lead to increased agricultural productivity because other factors also matter. It can also lead to labour displacement with social and political consequences (Bandyopadhyaya, 1971). Moreover, urban–rural integration should not be pushed by those wanting land enclosures for large development programmes. Rather, it should serve peasants development needs through sound interactions between urban and rural in terms of the former nurturing the latter with preferential inflow of resources to stimulate the rural economy. It is of crucial importance that China should avoid the likelihood of being trapped in its urbanization process, as seen in India and Latin America, marked by rising inequality between poor and rich urban dwellers.

Conclusion

Whereas land shareholding cooperative organizations are proclaimed to be the solution to various agricultural challenges, especially land fragmentation often coupled with land waste as a result of outmigration, there has not been a systematic and detailed survey on the effectiveness of these organizations from a pro-poor angle. According to limited preliminary findings, for instance, even in Chengdu and Chongqing, which are on the spotlight of progressive institutional transformation setting good examples for the rest of China in implementing the urban–rural integration agenda, land shareholding cooperatives are found to be used as an umbrella for illegal land usages whereby farmland was turned into construction land for real estate developments. Lack of respect for peasant rights and the incidences of forced land transfer are not uncommonly seen in many local experiments (Deng and Huang, 2009). Land shareholding cooperatives reflect the fact that any attempt to advocate a uniform system of land tenure in rural China may fail due to China's immense diversity (Kung, 2000).

The development of land shareholding cooperatives reveals the fact that peasant shareholders do not have sufficient power to decide on how their land ought to be utilized, and how the benefits from the land can be distributed to them. Similar to the issue of the village collective, questions remain as to who exactly represents the cooperatives, how individual shareholders can exercise their rights, how they reap the benefits, and to what extent they can have their say in management. In the absence of adequate shareholder participation, it is hard to tell how the cooperatives have served the best interests of the peasant shareholders. Lucrative deals can be made between village cadres and corporations in land management behind closed doors (Cai, 2003; Po, 2008).

Addressing the issue ofthe abuse of power by the village administrative committee and higher-level government in controlling land operation and governance as a threat to the ongoing reform (see Guo, 2001; Cai, 2003) is never an easy task in view of the multi-functional nature of land tenure interwoven with complex social, political and economic relations. Although market reform is inextricably linked with clarified and strengthened individual property rights, this does not necessarily mean that collective action or collectively defined property rights are irrelevant. Therefore, a land shareholding system is not inherently unworkable if it is based on the free will of the peasant shareholders. Essentially, it is all about how it serves the interests of the poor whose land ought to be utilized and managed optimally so as to achieve sustainable development.

The land shareholding cooperative is thought to contradict the merits of the HRS, since it is attributable to the abuse of the financial and democratic power of the peasants by the larger shareholders such as business enterprises and even village collectives. Thus, in certain cases it has disempowered the poor shareholders rather than empower them as originally designed and claimed (Kung, 2003). Yet, many government officials believe that it is not aimed at changing the nature of the HRS and replacing it. Rather, it is the modification of the system that suits the trajectory of economic reform. In practice, however, the role of land shareholding cooperatives in agricultural development should not be overstated, as agricultural development relies upon comprehensive support of the government which has to deal with structural limitations of agriculture. One of its major risks concerns the loss of land of the poor whose migration to cities can terminate the guarantee of land as a basic means of social security especially when their rights and access to social security are not adequately provided in the cities (Hu, 2009).

The land shareholding cooperative system represents a hybrid form of capitalist shareholding and socialist cooperative institutions. There exists a considerable divergence between original policy intentions and the eventual outcomes of the cooperatives. Sound management of the land shareholding cooperatives to strengthen the existing rights of the peasant shareholders is absolutely needed.

Nevertheless, the enforcement of peasant land rights can be undercut by many complex political, economic and social factors. Moreover, local authorities may perceive this as a threat to their power. They would have fewer incentives to implement the relevant laws and policies properly. And they may seek last resort measures to intervene in the ways in which the peasants' land rights are delimited, determined, used and managed in the name of land consolidation for the purpose of scaled-up agricultural production through the institution of cooperatives. The development of the land shareholding cooperative system may serve the economic interests of the local state and corporations far more than the livelihood needs of the peasant shareholders.

The current land shareholding cooperative system is not a panacea to solve the underlying issues concerning efficient and optimized land use in China. Peasants' reluctance to vacate their land to facilitate the development of the land lease markets in shareholding arrangements indicates a certain level of market and institutional failure. Furthering the market-oriented land tenure reform without addressing the key governance issues will not improve the current situation. The arguments developed here run counter to those who believe that the market-oriented system can break down the institutional barriers (see Yao, 2000). Nonetheless, the institution of land shareholding cooperatives presents an alternative to land privatization that is not feasible within the current political, economic and social parameters. It has the potential to demonstrate how the complex relations among different stakeholders can be ideally re-formulated for the sake of intensive and efficient land use that can benefit all, especially poor small landholders. It is a further indication of the aim of the government economic reform to integrate rural and urban development characterized by equal rights and opportunities for both rural and urban residents.

One needs to understand the ongoing village democratization processes in which peasants' rights need to be significantly improved so that they can wield more power to participate in decision-making processes concerning land-related institutional development. Policy-makers and theorists ought to avoid using the institution of land shareholding cooperative as a prototype of land reform without allowing for community-centred approaches to institutional innovation that better suit their needs in the local context (see Banks, 2003). Further studies of the impact of the land shareholding cooperatives on the members' rights, livelihoods and agricultural development as well as member organizations and their responses to the institutional changes are required.

References

Bandyopadhyaya, Kalyani. 1971. "Collectivization of Chinese agriculture: Triumphs and tragedies (1953–1957)", *China Report*, 7, 42–53.

Banks, Tony. 2003. "Property rights reform in rangeland China: Dilemmas on the road to the household ranch", *World Development*, 31 (12), 2129–42.

Bramall, Chris and Jones, Marion. 2000. "The fate of the Chinese peasantry since 1978", in Deborah Bryceson et al. (eds), *Disappearing Peasantries? Rural Labour in Africa, Asia and Latin America*, London: Intermediate Technology Publications.

Cai, Yongshun. 2003. "Collective ownership or cadres' ownership? The non-agricultural use of farmland in China", *The China Quarterly*, 175, 666–70.

Cao, Jinqing. 2005. "中国农民最需要什么" (What the Chinese Peasants Want the Most), http://www.nongyanhui.com/showart.asp?id=59, accessed 23 June 2009.

Cheng, Nianqi. 2006. 国家力量与中国经济的历史变迁 (*State Power and Historical Change of the Chinese Economy*), Beijing: Xinxing Chubanshe.

Chi, Fulin. 2010. "让农民工成为历史: 十二五推进城乡一体化的重大任务" (Letting rural migrants become history: Key tasks to promoting the integrated rural–urban development during the 12th Five-Year Plan), China Institute for Reform and Development (ed.), *Trends and Challenges of Urban–rural Integration in the 12th Five-Year Programme Period*, Beijing: Zhongguo Chang'an Chupanshe.

China Institute for Reform and Development (CIRD) (ed.). 2010. *Trends and Challenges of Urban–rural Integration in the 12th Five-Year Programme Period*, Beijing: Zhongguo Chang'an Chupanshe.

Deininger, Klaus and Jin, Songqing. 2009. "Securing property rights in transition: Lessons from implementation of China's rural land contracting law", *Journal of Economic Behaviour and Organization*, 70, 22–38.

Deng, Li and Huang, Wen. 2009. "成渝两市农村土地股份合作组织现存问题及完善措施" (Sichuan and Chongqing City Rural Land Shareholding Cooperative Organizational Issues and Measures For Improvements), http://www.papershome.com/economic/place/5406.html, accessed 15 December 2009.

Dongfang Zaobao. 2008. "广东: 土地流转改革一度遭批斗" (Guangdong: Land Transfer Reform Under Attack), 12 November 2008.

Guo, Xiaolin. 2001. "Land expropriation and rural conflicts", *The China Quarterly*, 166, 426–7.

Hinton, William. 1994. "Mao, rural development, and two-line struggle", *Monthly Review*, 45 (9), 1.

Ho, Peter. 2005. *Institutions in Transition: Land Ownership, Property Rights and Social Conflict in China*, Oxford: Oxford University Press.

Hoff, Karla and Stigliz Joseph. 2004. "After the big bang? Obstacles to the emergence of the rule of law in post-communist societies", *American Economic Review*, 94, 753–63.

Hu, Jing. 2009. "A critique of Chongqing's new 'land reform'", *China Left Review*, http://chinaleftreview.org/index.php?id=56, accessed 26 December 2009.

Hu, Yamei, Huang, Zuhui, Hendrikse, George, Xu, Xuchu. 2007. "Organization and strategy of farmer specialized cooperatives in China", in Gerard Cliquet et al. (eds), *Economics and Management of Networks: Franchising Strategic Alliances and Cooperatives*, Heidelberg: Physica-Verlag.

Huang, Yanhua. 2005. "广东南海: 农村股份合作制触到三个边界" (Guangdong Nanhai: rural shareholding cooperative's three frontline issues), *China Reform*, http://news.xinhuanet.com/report/2005-07/11/content_3203793.htm, accessed 5 December 2009.

Jiang, Xingsan and Liu, Shouying. 2004. "土地资本化与农村工业化: 广东省佛山市南海经济发展调查" (Land capitalization and rural industrialization: An investigation of the economic development of Nanhai, Foshan, Guangdong Province), *Economics Quarterly*, 14, 211–28.

Knight, John and Song, Lina. 1993. "The spatial contribution to income inequality in rural China", *Cambridge Journal of Economics*, 17, 195–213.

Kung, James Kai-Sing. 2000. "Common property rights and land reallocations in rural China: Evidence from a village survey", *World Development*, 28 (4), 701–19.

Kung, James Kai-Sing. 2003. "The role of property rights in China's rural reforms and development: A review of facts and issues", in So, Alvin Y. (ed.), *China's Development Miracle: Origins, Transformations and Challenges*, New York: M.E. Sharpe.

Li, Mingsan. 2008. "在集体框架下推农村土地股份合作制" (Promoting rural land shareholding system under the collective land ownership framework), 21st Century Economic Report, 7 October 2008.

Nanhai Agricultural Bureau. 2007. *Nanhai Agricultural Development: The 11th Five Year Plan and Longer Term Plan*, http://www.nhagri.com/survey.php?ncid=18, accessed 5 December 2009.

Plummer, Janelle. 2004. *Community Participation in China: Issues and Processes for Capacity Building*, London: Earthscan.

Po, Lanchih. 2008. "Redefining rural collectives in China: Land conversion and the emergence of rural shareholding cooperatives", *Urban Studies*, 45 (8), 1603–23.

Qin, Hui. 2006. "地权归农会不会促进土地兼并?" (Land to the tillers: Causes of land concentration), 经济观察报 (*Economic Observation News*), 21 August 2006.

Rozelle, Scott. 1994. "Decision-making in China's rural economy", *China Quarterly*, 137, 99–124.

Skinner, G. William. 1965. *Marketing and Social Structure in Rural China*, Ann Arbor: Association for Asian Studies.

Wang, Zhuo and Xu, Bing. 1996. 中国农村土地产权制度论 (*A Theoretical Study on China's Rural Land Property Rights System*), Beijing: Economic Management Press.

Wen, Tiejun. 2004. "耕地为什么不能私有化?" (Why can't the farmland be privatized?), China's Reform, Issue 4, http://business.sohu.com/2004/05/09/70/article220057016.shtml, accessed 3 December 2009.

Williamson, Oliver. 2000. "The new institutional economics: Taking stock, looking ahead", *Journal of Economic Literature*, 38, 595–613.

Yao, Yang. 2000. "Land lease market in rural China", *Land Economics*, 252–66.

Zhang, Hongliang. 2009. "土地股份制沉没前的最后一块木板" (The Last Moment of The Land Shareholding Cooperative System), http://bbs.cnnb.com/thread-1033227-1-1.html, accessed 3 December 2009.

Zhang, Qian Forrest and Donaldson, John A. 2008. "The rise of agrarian capitalism with Chinese characteristics: agricultural modernization, agribusinesses and collective land rights", *The China Journal*, 60, 25–47.

Zhou, Kate Xiao. 1996. *How the Farmers Changed China*, Boulder: Westview.

Zhao, Weiqing. 2007. 浙江省农村社区（土地）股份合作制改革问题研究 (*Study of Rural Land Shareholding Cooperatives in Zhejiang province*), Hangzhou: Zhejiang People's Press.

Zweig, David. 2000. The 'externalities of development': Can new political institutions manage rural conflict? In Elizabeth Perry and Mark Selden (eds), *Chinese Society: Change, Conflict and Resistance*, London: Routledge.

Chapter 7

Innovative Pro-poor Collective Land Tenure: Case Study of a Village Commune in Southern China

Introduction

Chapters 5 and 6 have discussed both individual and collective land tenure systems that reflect the mainstream policy developments and local practices concerning rural land tenure reform and its linkages with governance and development. It is important to note that the so-called market-oriented individualistic land tenure and shareholding arrangements are not contradictory but complimentary. On the one hand, more strengthened individual peasant households' land and property rights are deemed essential in safeguarding their interests and assurance of tenure security. On the other hand, this mechanism is assumed to lead to the transparent operation of the land shareholding cooperatives, as individuals can use rights to hold the cooperative management accountable. However, it has been shown that both systems have brought about unintended consequences of social conflicts, economic marginalization of weaker groups and land resource misuses, as the ostensible 'rights-based' approach diverges from its underlying context.

In the 1950s–60s, state-led agricultural collectivization, especially represented by rural people's communes, came under fire because of their failure to generate increased agricultural production and peasant incomes. Interestingly, the justification of the commune was based on the state's aim to tackle the structural problems of rural inequality – differences between rich, middle and poor peasants, and in particular, the reality of small landholdings which were identified as hampering agricultural organization and mechanization (Bandyopadhyaya, 1971). The same objective behind the current land policy changes appears to address the issue of land fragmentation, but in practice in certain cases obstructs more efficient farming and marketing of agricultural produce, as illustrated in Chapter 6. This chapter offers a unique example of a shift away from state-led approaches to locally-devised land tenure. Regional economic, social, political and bio-physical differences require diversified solutions to rural development and land tenure regimes. There is a commonly-held view that a successful land policy cannot be designed at the national level, but must cater for the possibilities and limitations of particular environments. Furthermore, land policy should aim at improving the livelihoods of the majority poor, while creating a viable basis for production growth and sustainable land use (Zoomers and van der Haar, 2000: 70). Despite the stronghold of the state in prescribing its institutions to the peasantry, local reluctance to adopt these changes and to cling to its own preferred modes of land use and management do exist in the Chinese countryside. This chapter's focus on a village commune reveals why and how the communal system has existed and confronted the mainstream market economy. It has been proven to be more effective in equitable rural development and in safeguarding the best interests of its members than state-led approaches. However, its recent development concerning the likelihood for its demise reveals the challenges of sustainable development in general and agrarian economy in particular, in the face of rural–urban transformation underpinned by the entrenchment of the market economy.

The key to successful rural land institutional development, to a large extent, hinges on the mechanisms to bring about stable economic development and rural governance. This chapter demonstrates that peasant deliberation and choice over the persistence of the commune system support this thesis. Any type of land tenure, be it individual ownership, shareholding cooperatives, or communal ownership, is a manifestation of local state–society interactions. This would require strong village governance with innovative design of land institutions to cope with political, economic and social constraints. In the context of mounting social inequality and rural poverty in China, the institution of the revitalized commune further sheds light on the role of collective action in overcoming these social dilemmas (see Ostrom, 2005). Ironically, with the recent retirement of the village party secretary, the village commune in this case study has to face a hard choice over whether the system should be continued or whether the land should simply be transferred or 'sold' to business ventures.

In the context of agrarian reform in many developing countries, the role of customary institutions has been controversial. Despite their positive contribution to land management in terms of greater space for social equity than modern institutions of individual land holdings, customary institutions are often criticized for entrenched unaccountability and even corrupt land management resulting from ongoing land administrative reform. To a large extent, village leadership or chieftaincy is often claimed to be critically responsible for elite capture and its associated impact on the poor in land reform processes (Ubink and Quan, 2008).

By contrast, the case of the village commune in this study shows that the village leadership is crucial to democratic governance and economic development, which does not necessarily follow government dictates. Its capacity to keep the commune intact from state intervention is revealing. It is important to note that, as compared with other country cases and even the old Chinese communes in the 1960s, the current commune displays a hybrid system in which both communal and market-oriented approaches to land management such as land leasing and shareholding cooperation co-exist and reinforce each other. Moreover, the village leadership, in the use of its power to define the communal rules, derives its legitimacy in serving the needs of the poor rather than its own interests as seen in many other cases (see Chanock, 1985; Firmin-Sellers, 1995; Oomen, 2002). Thus the economic, social and political dynamics of this system determine its sustainability and value in rural development and land management. However, the sustainability of this system is put at risk, as village power balances shift in the face of the broader spectrum of development that has not adequately addressed the pressing need for sustainable land use and agriculture. As a result, conflicting interests in land among peasants and other stakeholders have become paramount, which challenges the communal leadership that has yet to forge stronger alliances with the peasants and the local state.

Regional Political Economy and Land Institutions

This case study was conducted in the Yakou Village of Nanlang Township, Zhongshan Municipality situated in the Pearl River Delta – the most developed economic region in southeastern Guangdong province – the richest region in China (see Figure 7.1). Yakou is claimed to be one of the few commune villages left in China.[1] Guangdong province was ranked the third most populated region among China's 31 sub-national economies. It is the province where the first parcel of land was 'sold' to foreign investors in 1987, and where many so-called innovative rural land use arrangements were experimented with in response to growing market demand for land (Prosterman et al., 1998; Miao, 2003). As elsewhere in the country, land expropriation by the local governments and village administrative committees (village collectives) for housing and infrastructural development is commonplace. During the period 1996–2004, agricultural land shrank drastically. Accordingly, land used for industrial and urban development expanded by 19 per cent, and land used for transportation increased by 40 per cent (Lin and Ho, 2005).

The alliance between the state and business enterprises has dominated the development process in which peasant and state properties such as land can be easily expropriated. Rampant conversion of agricultural land to non-agricultural uses has exerted an adverse impact on the poor peasants' livelihoods and sustainable land use (de Angelis 2001; Li, 2006). Rising landlessness among the poor is also coupled with widening social inequality in this province over the last two decades (Fewsmith, 2007).

In Guangdong, land policy responses have displayed certain features of institutional innovation that allow for more market-oriented approaches to land transfers deemed necessary by the local government to protect peasants' land rights, for instance in the form of land shareholding arrangements (see Chapter 6). Despite these efforts in land transfer processes, peasants have little power to resort to the law to defend their interests. According to the Provincial Land Department, this is exemplified by the lack of appropriate stipulations on peasant land rights

---

1   The exact number of commune villages in China is unknown, but there are a few spread out in other provinces such as Jiangsu, Henan, Hebei and Zhejiang. Yakou village is called the last People's Commune in China given the fact that key features of the commune of the 1960s are maintained (Nanfang Daily, 2008).

Figure 7.1   Yakou Village, Guangdong Province, China

in law, which is exacerbated by the lack of coordination and clear division of responsibilities among its line agencies. For instance, the relationships between the land, construction, forestry and water departments are often blurred. Their conflicting interests can cause increased land administration costs and exclusion of the peasants in land governance.[2]

The registration of rural collective land use rights aimed at clarifying and protecting individual household land use rights is a priority for the Guangdong provincial government as elsewhere in China. In practice, rural land registration is not based at the level of individual households (except for housing land); rather, the entire village remains as the basic unit of registration. Only in rare cases are land registration certificates issued at the natural village level to resolve disputes and avoid conflicts over village boundaries and unsettled claims due to a lack of technological equipment and skills in land survey and cadastral management. As such, rural land registration mainly serves the needs of land administration

_____

2   Interviews with Guangdong Provincial Department of Land and Resources officials in May 2008.

or policy mandates rather than addressing the complexity of land relations and protection of peasant land rights.[3] In short, to a certain extent, it remains an empty institution and has not proven to be effective in improving land management to settle various land disputes and historical claims whereby peasants' interests in land can be safeguarded (Ho, 2005).

The provincial government does not favour individual land tenure or the HRS because the latter obstructs economies of scale in agricultural production. More collectively-managed land use is desirable in this province and other relatively developed regions where farmland production has given way to land investments, as a result of which rural–urban migration has become paramount. Moreover, in these regions, peasants' legal awareness and ability to participate in village governance are more developed than those in poorer regions. Thus, they have more capacity to oversee land management processes. The more developed a region is, the better the rural land is collectively used and managed, which can facilitate smooth land transfers. By contrast, in remote poor areas the HRS should be upheld, as the land itself has not much market value and can still be in the hands of the contracted households who rely on the land to conduct their traditional way of life.[4]

Nonetheless, a collectively-managed land use system favoured by the government does not necessarily equate to the one as peasant-centred and self-organized. Because of a lack of consensus on the type of land rights and management that best suit the needs of the peasantry, local governments and businesses, land institutions may favour the local economic development imperatives more than the needs of the peasantry. The disjuncture between land policies and local realities has continued to complicate the way in which land use and management is carried out. The following account shows how the village collective has responded to the policies and how they have used the land communally to maximize their collective interests especially for the weaker groups in the village.

Local Responses to Land Institutional Reform

Yakou Village seems to have been insulated from the mainstream political economy especially with regard to its own village development patterns. It is among the top 10 most beautiful villages in the province with a large potential for tourism. With a total population of 3,131 and 928 individual households, it has eight sub-villages consisting of 13 groups or production teams. It is rich in natural resources such as fertile soil, water and forests. It has 3,000 *mu* (15 *mu* = 1 ha) of land for rice

---

3   Interviews with Guangdong Provincial Department of Land and Resources officials and experts of the China Land Survey and Planning Institute in May and July 2008 respectively.

4   Interviews with Guangdong Provincial Department of Land and Resources officials in May 2008.

cultivation and over 20,000 *mu* of tidal land that has been developed for fisheries over the last 20 years. Above all, its farmland has always been kept under the management of the village commune.

In China, with the nationwide implementation of the HRS to replace the rural people's communes in the late 1970s coupled with the market reform, the early period of 1978–1984 saw dramatic increases in annual rural incomes of 15 per cent per year. But since then, the Chinese peasants have encountered multiple difficulties, which show that the HRS has not functioned to the degree originally envisioned. Increases in peasant incomes began to slow down, contract and in some regions even reverse (Hart-Landsberg and Burkett, 2004). Moreover, in recent years, food security has been put at the top of the political agenda due to the fact that by 2004 China's agricultural trade deficit ran high as a result of a jump in imports. With China's entry into the World Trade Organization (WTO), the Chinese peasants have experienced negative impacts on their livelihoods and become more vulnerable to the volatile economy at local, national and international levels.

Dating back to the late 1970s, Yakou peasants faced a difficult choice as to whether the HRS should replace the commune. At that time, they had experienced the trend of outmigration as farming gradually became less rewarding for many, while off-farm opportunities provided a supplement. The majority of those who remained were not capable labourers. If they had followed the HRS, they would not have been able to till the land as efficiently as possible due to a lack of capabilities and mutual support, according to the village administrative committee. The village Party Secretary, who had been in power since 1974, strongly believes that the HRS fragments rural relations and community cohesion and undermines the village capacity in pursuing collective solutions to human and biophysical problems. He asserts that some village assets like the land should not be distributed to individual households, whose diverse interests would not be easily compromised. Land quality differs from plot to plot, which would only result in conflicts if it is individually owned. And its fragmentation would further lead to peasants' vulnerability to natural disasters. In addition, the HRS would facilitate farmland conversion in the hands of the state and businesses, for individual land users cannot challenge the power of the state in land expropriation, and this would be a disaster for the landless poor. After heated discussions within the village, a referendum was held and a consensus on the continuation of the commune system was reached with a view to protecting the most vulnerable groups especially the elderly and disabled. The decision to continue the commune system was luckily acquiesced in by the then local government.[5]

In the 1980s, plans for industrial development to trigger rapid village economic development were tried out. Despite the attainment of limited positive outcomes, there was an increased uncertainty over profit-making and capital costs. The village administrative committee soon realized that they would shoulder more risks in

---

5   This account was derived from interviews with the village Party Secretary and other village leaders in June 2008.

operating factories than sustaining agriculture, as the former had a more negative impact on the natural environment and in the long-run would be unsustainable. Thus, they decided to close down a number of factories and to shift their development priority to land leasing (Wang, 2007). Moreover, they understood that China had two social welfare systems for urban dwellers and the peasants, partly due to the HRS (Perry and Selden, 2003). The peasants have nothing to rely on but their land to meet their needs for livelihood and social security. By fully developing the agricultural potential of their village, they believed that the double goal of agricultural development and social protection of the commune members could be materialized (Yakou Village Administrative Committee, 2005).

Land Resource Management in Yakou and its Adjacent Villages

Land for Rural Enterprise Development in Yakou

Yakou peasants wisely utilize its 60,000 *mu* of farmland annually and have attained the 'xiaokang' standard (enough to eat and live on) since 2000.[6] This is demonstrated by the fact that 10 per cent of the households own private cars and 95 per cent live in modern houses built in the 1990s – an outstanding achievement largely due to the way in which the village is governed and how its endowed natural resources for the development of fishery and paddy rice farming are managed.

As land is treated as an invaluable asset by the village leadership, over the last 10 years Yakou peasants have turned the sand deposited by tidal waves into a cultivable area. This means that they have reclaimed the land from the sea and extended their coastline by 2.5 kilometres (Nanfang Daily, 2008). Given the fact that many local peasants do not have the know-how to utilize the reclaimed fertile land economically, the village committee promptly decided to take advantage of the market to lease this land to some peasants from the neighbouring provinces to develop the fishery sector.

In order to maximize the profits from the tidal land, in 2002 the village committee established the Farmland Shareholding Foundation responsible for the management of land leases, land rent collection and distribution to its shareholders. It is open to all the households and enables especially women and the elderly to become land shareholders. Land shares can be inherited but not transferred to people outside the village in order to ensure management integrity. It stipulates that for the tidal land that falls within the domain of the foundation, each shareholder maintains an average of 5.5 *mu* that constitutes their land share and the basis of dividend distribution – a major step to ensure equity in land share and dividend distributions (see Table 7.1). As a result, up to 2007 there had been an annual increase of RMB 500–800 in each shareholder's income. This figure is

---

6   The *Xiaokang* standard is the long-term development goal of the Chinese government, as average Chinese citizens have yet to attain this standard.

Table 7.1   Yakou tidal land shares distribution

Sub-villages	Number of households	Population	Number of males	Number of females	Number of land shares to each peasant (*mu*)	Total land area assigned to each sub-village (*mu*)
Dongbao	150	520	273	247	5.5	2,860
Pingshan	66	231	112	119	5.5	1,270.5
Yangjia	124	435	225	210	5.5	2,392.5
Zhongbao	153	5,267	2,534	2733	5.5	28,968.5
Huamei	52	176	97	79	5.5	968
Xiangxi	120	382	172	210	5.5	2,101
Lujia	95	345	170	175	5.5	1,897.5
Xibao	156	511	252	259	5.5	2,810.5
Total	916	7,867	3,835	4,032	5.5	43,268.5

Source: Yakou Village Administrative Committee, 2005, *Yakou Village Record*, p. 69.

expected to substantially increase as the gains from investments materialize in the coming years.

Non-agricultural land is also contracted to factories for the purpose of rent collection, which constitutes another major source of revenue for the village. Contrary to direct operation of these types of enterprises in the 1980s, the village administrative committee upholds the principle that the village itself is not involved in direct manufacturing to avoid capital and management costs. Moreover, no large-scale industries are allowed in the village in order to prevent air and water pollution, and all investments must undergo a preliminary check to ensure that they meet environmental protection standards set by the committee. Land management is underpinned by transparent governance to ensure peasant participation in decision-making over land contracting matters. Decisions over the approval of these investments are based on the consensus reached by the two-thirds majority of the committee members, as well as the representatives of each sub-village.

In short, the utilization and management of the tidal and non-agricultural land under market mechanisms are aimed at providing the peasants with the means of better livelihoods given the unique advantages of the land resources that carry great economic value for quick profit. In the context of the rural–urban divide in China and in terms of social inequality between urban and rural residents, sustainable land utilization for the maximization of economic benefits is seen as fundamental to the Yakou peasantry whose lack of employment and other economic opportunities make the land an invaluable asset for their livelihoods (Yakou Village Administrative Committee, 2005).

Land for Communal Agriculture in Yakou

The Village Administrative Committee chaired by the Party Secretary Lu Hanman strongly believes that the farmland is the backbone of rural life and should never be sold or used in any other way. This is a fundamental reason for the continuity of the commune system whereby the farmland has never been contracted to individual households and outsiders except for those plots acquired by the state over the last 20 years. Furthermore, it was the collective action marking the Chinese revolutionary success that the village leadership and majority members believe to be crucial to land equity. In the early 1950s, 83.7 per cent of the land was owned by the landlords, whilst the rest was owned by others (Yakou Village Administrative Committee, 2005: 77). To a large extent, under the commune system, the peasants have been able to sustain and substantially improve their livelihoods. Till now, 600 out of 1,700 capable labourers are commune members who engage themselves in rice farming covering 35,000 *mu* of land. The commune members are either the elderly who lack education and skills suitable for urban employment, or those returned migrants whose employments in the cities were terminated and had to seek shelter in their village of origin.

Under the current system based on the principle of 'collectively-organized labour contribution and equitable income distribution', the organization of rice farming is based on three levels – the administrative village (often called brigade), sub-village (production team) and individual households. Each production team is accountable to the brigade responsible for target-setting, technical support and oversight of production. Division of labour depends on demographic differences and no compulsory tasks are given to the members. In contrast to the old commune system in the 1960s, labour inputs are directly linked to the final distribution of rice harvests. A system to record individual labour inputs called *gongfenzhi* (system of work points) was inherited from the past with some modifications to ensure the accuracy of each worker's time spent in the field. The work points are calculated in accordance with the tasks of individual members who are free to decide on the time needed. For instance, if one has chores at home, he or she may not perform as much as work in the field as those with more time for rice farming. Upon each harvest, each production team then works out the incomes of each member based on the total income of the team and members' labour inputs. Equitable distribution of remuneration for farm labour is thus guaranteed. This practice differs from the commune in the 1960s when everyone could enjoy *daguofan* (eating food together from the same bowl). Moreover, the brigade purchases the grain from each production team at a price of 50 per cent above the market value and then sells it to the members at 30 per cent lower than the market price. Since 2001, children below 16 years of age and elderly peasants above 60 years (for women it is 55 years) and disabled groups have enjoyed a grain ration free of charge. After all, only a small proportion of the grain is sold on the market. The commune adopts this rule to deal with unexpected food insecurity and to maximize the members' benefits and interests in farming.

Further rules on helping the vulnerable poor are put in place. For instance, the peasants who migrate to cities can always return to the village and become commune members to till the land, in case they encounter unemployment in the cities. In the event of return and their wish to work in the field, they are required to seek the village's approval in the beginning of the year and to pay a fee of RMB 100. Once they have been regrouped into the production team and started farming, they cannot move out in the same year to ensure that farming is not affected. Although they seem to have limited freedom, all these measures aim to keep the balance in the overall agricultural production and demographic changes to ensure that the commune runs smoothly. But those who are unwilling to work in the field do not enjoy this treatment. Thus, a system of equality is ensured that is open to every peasant. No one is forced to take part in any communally-organized activities.

The commune members point out that this system enables scaled agricultural production. To the village leadership, organized farming and industrial development is essential to the provision of social welfare for the peasants. In fact, few profits are derived from rice farming per se, which is subsidized with the revenues from the collection of tidal land and other non-agricultural land rents.

Till now, there has been an annual income of RMB 10 million derived from this rent, RMB 6–7 million of which is used as agricultural subsidy. In other words, the commune system has two sub-systems – a communal system subsidized by a market economy as Lu indicated (Cao, 2002, 2004).

In addition, with the well-developed social welfare system including provision of housing, clinics, pension and special care for the elderly, few Yakou peasants wish to move to the cities. The women especially are not willing to apply for urban residency when they marry urban residents. By doing so, they can still retain their social welfare benefits. As some informants revealed, becoming urban residents mean that they would become 'hungry residents' afterwards. And the function of the land, especially the farmland, embodies more value of social protection than agricultural production.

Land for Enterprise Development in the Neighbouring Villages

In stark contrast to the case of Yakou, the neighbouring Cuiheng village is a case in point with regard to farmland loss and aggravation of peasants' livelihoods. According to the Cuiheng village administrative committee,[7] the average household annual income is in the region of 6,000 Yuan, which is far below the Yakou standard. Yakou and Cuiheng used to be one village, but split after the rural administrative reform in 1998. Since the 1990s, Cuiheng has experienced state land expropriation at an alarming rate. To date, tens of thousands *mu* of farmland have been converted to construction land for real estate and industrial development. However, much of the expropriated land remains either underutilized or idle. The affected peasants were given the choice to buy new houses; however, with very limited compensation received, they could not afford them but had to move to other places. The local government has begun to redress these issues since 2007 by improving land expropriation procedures including issuing land use certificates to house owners and farmland users to safeguard their property rights. Yet, not more than a few hundred *mu* of farmland are left in Cuiheng, as most peasants have either migrated to the cities or stayed in the village and been involved in non-farming activities. The Village Administrative Committee has made use of the remaining land for industrial development to raise some revenue. Besides, they also induced peasants from other regions to work on the land taking advantage of the latter's skills and techniques in growing cash crops.

Home to Sun Yat-sen, founder of the Republic prior to the People's Republic, Cuiheng is known for its history. The local government has strived to make it a tourist destination as well as an area for business development. Two instances of development are prominent examples to demonstrate the impact of new land use development on the livelihoods of the poor and responses of the local peasants and village committee.

---

7   Interviews with Cuiheng Village Administrative Committee members in June 2008.

In the first instance, the establishment of 300 *mu* of Zhongshan Movie and Television Town was a huge project to showcase the history of the village and Sun Yat-sen. It also has hotel and entertainment venues. In order to exhibit the village architecture to tourists, a large proportion of the village houses have been kept intact in the style of the Qing dynasty, but the owners were asked to vacate their houses and land, which sounded odd to them in terms of the need for it. Most households have left for the cities and other villages with a few households having resisted vacating their homes. Compensation for the displaced households had not been largely agreed on especially for those who remained on the site

In the other case, in 2008 the expansion of a reputable secondary school also met with difficulties in obtaining the agreement from the affected households. The latter's concerns were mostly about the resettlement plan and the amount of compensation provided by the government. This project was listed as one of the major undertakings of Zhongshan municipality, since the school is under the jurisdiction of the Zhongshan Education Bureau. In a request or rather a demand to the municipal government, the school management put forward two issues constraining the school expansion process – delayed and incomplete household removals and 101 *mu* of orchard that had yet to be acquired. Immediately, the government gave the project a higher priority and convened a meeting to coordinate with different line agencies as well as the township government that had been blamed for being too slow in completing the project land use plan, and warned that further delay would bring about the failure of the school to admit new students in the next semester. All these pressures were placed on the Cuiheng village administrative committee that was obliged to accelerate its land acquisition process in which more peasants would be affected. Apart from the issue of insufficient compensation for the affected households, some did not want to vacate because they saw the value of their properties increasing, especially for those living in close vicinity to the market. 'Even if I am given a compensation fee of RMB 1 million, I still want to stay here. But it is not possible to win any deals with the government that can use any force including public security guards to force us to leave', as one informant pointed out.[8]

Overall, on average, each household received an estimated RMB 10,000 as land compensation without adequate social insurance guarantee. Although they were also provided new houses, they did not receive housing ownership certificates due to the unwillingness of the real estate agencies to apply for the certificates for them. And the cost of a few thousand RMB also deterred many from applying for the certificates. Likewise, another neighbouring village, called Xiasha, has suffered from land loss since the 1990s. In 1992, under the pressure of the local government, Xiasha had to vacate almost all the land except the 300 *mu* of land earmarked for building a new village for the evicted peasants. Although they were offered new houses, they only received an average of RMB 10,000–20,000 as

---

8   Fieldwork in Cuiheng village in June 2008. This section and the following paragraphs are based on interviews with the peasants and village leaders in Cuiheng.

compensation. At that time, this was seen as a big sum, which caused jealousy among some Yakou peasants. However, a large proportion of the expropriated land was left undeveloped, and it was no longer suitable for farming. Furthermore, many evicted peasants had left the village to seek off-farm opportunities in the cities and could not return to the village because there was no land left for them (Cao, 2004).

Conflicting Perspectives

Reflecting on the pressing land issues, the Cuiheng village leaders thought highly of the value of the land for village development as well as the role of government in guiding land development through scientific and integrated development plans rather than merely pushing peasants into the land market assumed to facilitate peasants' land transfers. More importantly, the land acquisition process must incorporate mechanisms for fair compensation to the peasants whose approval of the plan must be made a priority. Currently, 70–80 per cent of affected peasants must agree to the land acquisition and compensation plan as required by the land law. They also admitted that some peasants and even some village leaders were unwilling to embark on landed rural development (indirectly pointing to the case of Yakou), which can be attributed to their limited insight, knowledge and capacity. They pointed out that rural development in China had always been a huge challenge and there was little to learn from the experience of the past and from other villages, as conflicting interests of groups and individuals always pose huge difficulties to decision-makers. For instance, where land acquisition is concerned, peasants' interests differ hugely. Some do not want to vacate their land; but in Cuiheng, many elderly people do because of their age and little hope of prosperity from the land itself. They would rather rely on the income earned by their children in the cities. Thus, they think highly of education as the only means to leave the village. The divergent views of the peasants and village leaders over land use in terms of how land can be best utilized to benefit local development and the poor do not always support the upholding of farming as essential to village development.

The Cuiheng village leaders foresaw that land would be privatized sooner or later in China. They did not explain this but emphasized the unstoppable trend of land market reform. Yet how land privatization could be brought about and what the effects on the poor peasants would be remained a puzzle to them. When asked about the land registration progress which is supposed to play a role in clarifying landownership and protecting peasants' land rights, they did not think that it was important to land management and overall local development, for it was just an administrative formality. For them the government's priority should be placed on ensuring peasants' land rights in the land management process. In other words, provided that all the land acquisition procedures satisfy the needs of the peasants, the latter would be willing to give up the land eventually. Moreover, they pointed out the importance of land use planning, which should be further strengthened.

Failure to do so had much to do with the government abuse of power and the lack of sound management processes concerning farmland conversion.

The majority of the peasant informants in Cuiheng held the view that it would be better if they were allowed to sell their land directly in order to circumvent the intervention of the local government. First, they believed that they could not do anything about the future of the land in the face of forced removal by the government. They realized that the land would be given away to the government sooner or later, and some even argued that there would be no farmland left in Guangdong in the foreseeable future. Second, agriculture in Guangdong, as in most parts of the country, is not profitable at all, which is a stumbling block to their incentives in farmland investment. Third, they simply wanted to keep small land plots for their housing and other needs and to lease the land when they find off-farm employment opportunities. Above all, a lack of access to legal aid and other means of social support would add fuel to the burning tensions between them and the local state. This is compounded by the ineffectiveness of land law and policy that can be easily manipulated by the alliance of local government, business enterprises and village leaders. For example, the law only requires two-thirds of the villagers to agree to the land acquisition plan. Yet they knew that in most cases as long as the alliance members could strike a deal, the peasants themselves would be left with little leeway but to sign the agreement. In this sense, they even argued that at the very least land privatization might give them more secure rights than collective landownership. Surprisingly, when commenting on the neighbouring village Yakou, the Cuiheng village leaders hinted that the Yakou peasants would have been better off had the land been contracted out because rice farming was not profitable. This view also resonated among some Yakou peasants who were concerned about the sustainability of the commune itself.

By contrast, the Yakou village leaders pointed out that Cuiheng had nothing left but unfinished buildings and fragmented housing land. For instance, they revealed that most of the enterprises in Cuiheng had not secured profits after land expropriation. As a consequence, the Cuiheng peasants had become landless, which in turn could undermine their customs, language and kinship relations. The Yakou village leaders also uphold the commune as a model in the reform era, as an opposing force against industrial penetration into the farmland. They showed their discontent with the market reform as compared with the pre-reform era in which the people's commune was paramount. Neither did they understand the meaning of the Chinese revolution in respect of the current situation where peasants' land is forcibly taken away by the local government and developers. For instance, they criticized the model pursued in Dongguan – a small booming industrial city in the province whose development has driven many peasants off their land. As a result, the farmland has become a site for industries which caused air pollution and environmental damage to the surrounding natural resources. In their view, this type of rural development only benefits the government and businesses. Thus, the commune is an effective institution that offers protection to the peasants against farmland loss to the mighty few. But they recognized the role of the economy in sustaining the commune system. Yakou is unique in its geographical location and natural resources – the key to the local economy. The same type of commune could not be feasible in other regions. In any case, the land redistributed to them through the revolution of the Communist Party ought to be preserved rather than snatched by others. Thus, the development of the land market and even potential land privatization would negate the Communist Party's struggle and the meaning of the revolution. In the spirit of equitable development and collective management of the village, the Yakou peasants managed to donate RMB 110,000 to the Sichuan earthquake victims and their families in May 2008, while only a few thousand RMB was collected in the neighbouring villages such as Cuiheng.

The Commune Under Threat

To the Yakou village leadership, land has never been the means for short-term profit gains. On the contrary, land preservation and its sound management have been given the highest priority. In fact, 'never sell our land' has been one of their guiding principles evident in their worship of the 'god of land' in all households that regard it as the symbol of peace and prosperity. Yet, for the village leaders, maintaining the commune has never been an easy task given the mounting political and economic pressures discussed earlier. They have had to cope with internal discontent and even the intervention of the local government in their silent struggles.

In recent years, there have been several rifts among the peasants as to whether the commune system ought to be discontinued. Concerning land use, some contended that the paddy fields should be sold to outsiders so that they could use that money to do whatever they wanted. They also expressed their concerns over the commune's agricultural inefficiency. Especially for some young peasants, they would rather use the land for other more efficient purposes, because they have the opportunities to seek off-farm employment. Some even contended that there was nothing wrong with land privatization as long as equality and their benefits from the land are ensured. They felt that they had lost the rights to directly use the land, which the HRS would otherwise have granted them as in other parts of China. As some saw it, the peasants in Cuiheng were in a better position to gain meagre profits from land expropriation, but the Yakou peasants have lost this opportunity to a large extent. Moreover, the commune system was criticized by township government for lacking initiatives to adopt advanced technology and undertake innovative agribusiness activities. It was thought that the commune should be transformed into to a more effective institution that helps the poor members out of poverty.

By contrast, for the elderly and unemployed, the farmland provides a minimum safety net on which they can always rely. They were not in favour of land privatization or decollectivization that would cause economic disparity within the village and fragmented community relations. They also mentioned that only those who managed to find jobs outside the village would favour the HRS. Overall, the research shows that in 2008 about 70 per cent of the informants showed their contentment with the fact that land was still in the hands of the village. At the same time, they thought that neither the commune nor land privatization would be a weapon against illegal land evictions. They advocated more solutions to improve land use efficiency.

These views reflect the ongoing challenges for the village leadership to address. The village leaders argued strongly that any trend toward the discontinuation of the commune would be short-sighted, unwise and lacking thorough factual basis. They did not believe that any potential land 'sales' would contribute to the maximization of profits from new development opportunities given the vulnerability of the market economy and the lack of employment opportunities for the commune members. As the Party Secretary Lu refuted,

> Dismantling the commune by dividing up the land or even allowing for land transfers might generate quick incomes to the members, but they must not forget that when the money is used up, they will have not much left for themselves and their offspring who will have no land to rely on in the end.[9]

To the commune leaders, the only way to ensure community equity is to keep the land as it is. It is believed that in the era of economic uncertainty, especially with low economic returns from agriculture, the commune plays an essential role in the rural economy in regulating inter-household relations to safeguard their best interests. However, they also recognized that there were increased pressures on the paddy fields. The more people kept returning to the village from cities, the more difficulty it was for the commune to accommodate their needs.

Struggles over the perpetuation of the commune between the peasants and the commune leaders reached a climax when a few members formed the 'anti-corruption action group' publicly accusing the village committee especially the Party Secretary of corruption and abuse of power in 2001. This group also sent petition letters to both the Zhongshan municipal and provincial governments. Their accusations focused on two issues – land management transparency and continuation of the commune. This action aroused the serious attention of the Nanlang township government which dispatched four special task teams to the village to carry out household investigations and designated township auditors to examine the village financial accounts of 1997–2004. Based on the 752 household surveys (87 per cent of the total number of households) and the auditors' reports, these allegations were proven false. The findings showed that the Party Secretary Lu had enjoyed a high reputation among the commune members that explained the effectiveness of the village leadership in fighting poverty and transparent village governance. As a result, the village committee was cleared of all the charges,

---

9   Interview with the Yakou village Party Secretary in June 2008.

which enabled the commune system to continue (Yakou Village Administrative Committee, 2005: 164).

Moreover, village–state land struggles also take place in silent and sometimes undetectable ways. The village leadership has managed to deter many claims and deflect discontent over the commune. As a result, the local government could only resort to other means to use the village land. To promote tourism, for instance, it built roads surrounding the village to pave the way for the development of a forest park at a later stage in the hills owned by the commune. By doing so, it may claim its ownership after the park is built, since by then the village would probably find it extremely difficult to manage the mountain resources. Another example shows that the local government also uses its development policy to promote its own interests. As Yakou is prone to typhoons and flooding, mangrove trees were planted along the coast to prevent natural disasters. However, according to the local peasants, this practice may not be effective and moreover, it could affect the marine ecosystem. Again, this was seen as another step by the local state to extend its control over the village. Obviously, the village and local state do not always have the same development goals, which is a driving force behind the struggles over limited land resources and the power to control them.

These power struggles again are embedded in the implementation of land policies. Yakou village was claimed to be the only village without rural collective land registration by 2008 as required by the 2007 Property Law and mandates of the central and local governments, despite numerous local government notices and warnings issued. The village leaders had a strong reason to refuse land registration. Local government had no other tactics to solve this issue but to complain about the backwardness and stubbornness of village leadership who were not open to further discussions. The local government found it difficult to reach an agreement with the village leadership especially the village Party Secretary whose insistence on the commune system had averted many of their interventions. According to the township land bureau, the Yakou village committee was concerned about land acquisition and farmland loss to the local government in the event of land registration. This also reflects the fact that local government or any public oversight over the land use arrangements is ineffective. The township land bureau does not possess sufficient knowledge about how the farmland is leased and cultivated to determine whether they meet relevant policies and requirements. The village collective in managing the commune seems to be able to make the best use of the collective landownership vis-à-vis the HRS as a weapon against the arbitration of the local government with regard to land use.

Nonetheless, these struggles have not eased and have even regained momentum in the broader context of farmland acquisition for the so-called local and regional development. In fact, even within the village, the disparity in livelihoods between the commune members and those with urban employment complicates their divergent interests in the future of the commune system. In 2008, the Zhongshan Municipality made a further step towards the development of the eastern regions along the coast, which includes Yakou. As a result, Yakou's 9,536 *mu* of coastal land was demarcated for tourism and real estate development projects and thus fell under the expropriation plan of the government. In reaction, the Party Secretary Lu rejected this plan and through negotiations with the local government he managed to allow the local government to use this land through leasing for 70 years. Although this agreement did not render immediate huge compensation to the peasants, Lu believed that it would at least ensure their livelihoods for the next decades. As such, he insisted that the rent must accommodate annual inflation of consumer products and foodstuffs to maximize the social welfare function of the leased land. However, some peasant groups desired immediate cash returns from direct land acquisition in order to receive an average of RMB 140,000 in cash per household as compensation from the government. Their discontent led to the convening of the village assembly of 3,334 people on this matter in July 2008, the result of which was that only 28 people were on the side of the Party Secretary Lu. Later on, Lu refused to change the original agreement with the local government on the land lease, but failed to insist on his decision to hold up to the land. Eventually, the land was expropriated, but he refused to collect the compensated sum of cash. In early 2011, he resigned at the age of 72 after 37 years in power as the village Party Secretary (Southern Weekend, 2011).

The Commune as an Effective Governing Institution?

Since the resignation of Lu as Party Secretary, the future of the commune is put into larger question. Yet, this cannot nullify the fact that the institution of the commune has been a weapon of the weak to manage their land and tackle some of the economic and political challenges facing them. The Yakou commune manifests itself as a more effective community-centred collective institution in serving the interests of the members than the HRS. In the current Chinese context, sound rural land governance requires a strong village leadership through a reasonably established democratic governance system as exemplified by their daily management and elections in the case of Yakou. Lu is recognized as their rightful leader – someone who is not involved in corruption, who is self-disciplined and passionate about helping his fellow villagers and with a strong belief in the power of collective action in village development and governance. He is a sophisticated village leader regarded as the most important person for the commune. There is no single case in which he was involved in banquets or dinners with visitors and local government officials. He is thrifty – most of the time he goes barefoot and rides a shabby bicycle. More strikingly, the village committee work most of the time including holidays even during the Chinese Spring Festival, and usually till 9:00 p.m. To make everyone in the committee equal, their salaries are kept almost the same regardless of ranks. Moreover, they all have the power to approve any village policy documents and financial dossiers. Consequently, the village committee is so strong that it is seen by the members and even outsiders as a shining example for the rest of the country in terms of the dedication of the leadership to village economic and social development.

Yet, the endangered commune puts into question the direction of rural development and urbanization in the Chinese context as well as the needs of peasants and the landless (Southern Weekend, 2011). Essentially, local peasants want more from the commune in terms of more rapid and greater economic benefits from land management. The dominance of the local government in deciding how land should be used makes the peasants doubt the sustainability of the commune. Given this uncertainty, and at a loss as to how their land can be better managed to meet their pressing needs, the commune has no longer been appealing to many members as the last resort to securing their livelihoods. Enhanced village governance may not render a mere solution to the uncertainty over the commune. Although the village leaders have attempted to improve the mechanisms for accountability and transparency built into their daily work, they have yet to develop more effective means to cultivate peasants' incentives to participate in village governance.

Conclusion

Land institutional change in China's market reform and economic transition cannot be understood without paying attention to the conditions and dynamics of local contexts. The case of the Yakou commune explains the fact that the local peasantry can determine their own forms of land institutions that better suit their needs and local economic, political and environmental parameters. The institution of the commune is a paradox in the mainstream land institutional reform – the HRS and land shareholding cooperatives, underpinned by a market-oriented approach to land use and management have proven ineffective in addressing poverty and poor governance. It also explains that collective or communal land arrangements, on condition that village economic development suits the needs of the poor, do ensure land tenure security.

The case of Yakou commune differs from the rural people's communes prominent in the 1960s in three major aspects. First, under the People's Commune, commune members had no freedom to choose their work. In Yakou, they are absolutely free to decide on the opportunities that suit them, since it is an open system. Second, under the People's Commune, there was no individual economy; everything was organized by the commune as a collective. As Oi (1999) points out, the whole incentive framework was distorted by the ideology of the commune. In contrast, in Yakou, except for the paddy fields under collective operation, the rest of the resources are managed in the light of market mechanisms through land leases to other parties. In this sense, the village has a mixed economy which allows for the achievement of both economic efficiency and social equity for the disadvantaged groups. Third, under the People's Commune, village leaders were appointed by the commune. In Yakou, they are democratically elected and represent the interests of the majority voters (Cao,

2002). All these features indicate that collective choice over land management can achieve better economic outcomes when the collective institution is able to adapt to the demands of the market economy and peasants. Thus, the Yakou land system is hybrid in integrating both market and collective institutions. In facing economic uncertainty and lack of social protection for the rural poor in the wider economic context in China, it has provided a viable alternative for the majority of its members.

Land is a manifestation of economic and political power struggles among peasants, local government and other stakeholders in the Chinese countryside. Land rights can be understood in terms of who actually occupies or owns the land. However, as Zhang (2004) contends, land displays its inherent feature as a symbol of state power. This is easier to understand for the Chinese peasants who all know that their land does not really belong to them, although their long leasehold and use rights have been greatly enforced. Zhang argues that it is hardly meaningful to discuss the landownership issue – privately or publicly owned – in China. Rather, it is more useful to explore the underlying issues of dialectic relationships between land rights and power, which is important to understanding justice and equity concerning land rights.

The Yakou commune appears to be a symbol of village power in managing the land and in its struggle with the local state, which has far-reaching implications for China's land governance. As the local state has increasing power in decision-making, and despite numerous policies and laws of the central government to tackle poor land governance and unsustainable land uses, without limiting the overt power of the local state these policies and laws would be ineffective in addressing the mounting issues of land tenure and rural livelihoods. The government itself is still trapped in its transition because of poor governance (Pei, 2006). In this respect, the Yakou commune exemplifies an institutional innovation that galvanizes the collective power and resources to maximize the economic interests of the village. Moreover, as a social institution – an alternative to the HRS which puts individual households at the mercy of the market and state control, it has shown its strength in confronting the local state in its use of collective force based on peasants' participation in village governance and economic development.

Social and political relations in the rural setting always embody complex struggles over land tenure. The commune is an important institution for the members' articulation of collective identity and a means of dialogue between the powerless and powerful (Coombe, 1998). However, peasants' concerns and disagreement over the perpetuation of the commune reveal the urgency of tackling issues of rural development, improvement in governance transparency, accountability and peasant participation on both village and local government levels – not only by the village, but also the local government. Without this approach, the covert village–state and intra-village struggles will continue. In this sense, at least, the commune appears to be a weapon of the weak, though itself confronting internal and external challenges (see Scott 1985; Walker, 2008). Thus, an in-depth study of the commune can explain the many issues and dilemmas concerning peasant–state relations over land and local development, which has far-reaching implications for the understanding of collective versus individual action in rural land reform in China.

To a large extent, the endangered future of the commune is contingent upon the power of the local government and real estate agencies in persuading the villagers to accept land development projects for short-term gains (*Nanfang Daily*, 2008). Intra-village social and political divisions can actually complicate the communal land rights arrangements. In other words, communal rights are not as homogeneous as conceived by the proposition of the common property regime (see von Benda-Beckmann, 2006; Nagendra and Ostrom, 2008). The Yakou village leaders have to mitigate the conflicting interests of their own members to provide them with a more effective sustainable development framework. And they need to forge wider societal support for improved efficiency and better governance; above all, through further empowering the members in participating in decision-making and institution-building more effectively.

The institution of the Yakou commune marked by the hybrid land tenure systems is a manifestation of the use of flexible institutional arrangements to achieve the current levels of land protection and equitable village development as compared with its neighbouring villages. Communal land tenure in the case of Yakou offers a re-thinking of ongoing debates on land tenure with important implications for policy options in the course of China's transformation as well. It is in this process that more community-centred and flexible policy approaches ought to be sought by policy-makers. The effectiveness of a land tenure system is embedded within the overall pattern of land use, rural governance and development, among other parameters. Yakou peasants have managed to make these combined elements work in order to sustain their commune for decades, but the mounting challenges for its continuation ought to enable policy-makers to rethink how sustainable development goals can really be obtained from the lessons of the rise and downturn of this commune. The recent call of the Yakou villagers for land 'sales' is a testimony to the partial failure of government development policies at the village and regional levels.

References

Bandyopadhyaya, Kalyani. 1971. "Collectivization and Chinese agriculture: Triumphs and tragedies (1953–1957)", *China Report*, 7, 42–53.

Cao, Zhenghan. 2002. 信念，效率与制度变迁 – 广东省中山市崖口村公社制度研究 (*Belief, Efficiency and Institutional Change: Study on Commune System in Yakou Village, Zhangshan City, Guangdong Province*), Beijing: China Economy Press.

Cao, Zhenghan. 2004. 伶仃洋畔的村庄公社 – 崖口村的公社制度及其变迁 (*A commune village on the Lingding Coast – The Yakou village commune and its institutional change*), Beijing: China Economy Press.

Chanock, Martin. 1985. *Law, Custom and Social Order: The Colonial Experience in Malawi and Zambia*, Cambridge: Cambridge University Press.

Coombe, Rosemary. 1998. *The Cultural Life of Intellectual Properties*, Durham and London: Duke University Press.

de Angelis, Massimo. 2001. "Marx and primitive accumulation: The continuous character of capitalist 'enclosures'", *The Commoner*, September (2), http://www.thecommoner.org, accessed 18 November, 2008.

Fewsmith, Joseph. 2007. "Assessing social stability on the eve of the 17th party congress", *China Leadership Monitor*, 20, 1–24.

Firmin-Sellers, Kathryn. 1995. "The politics of property rights", *The American Political Science Review*, 89 (4), 867–81.

Hart-Landsberg, Martin and Burkett, Paul. 2004. "China and socialism: Market reforms and class struggles", *Monthly Review*, 56 (3), 7–123.

Ho, Peter. 2005. *Institutions in Transition: Land Ownership, Property Rights and Social Conflict in China,*. Oxford: Oxford University Press.

Li, Fan. 2006. "Unrest in China's countryside", *China Brief: The Jamestown Foundation*, 6 (2), 6–8.

Lin, George. and Ho, Samuel. 2005. "The state, land system, and land development processes in contemporary China", *Annals of the Association of American Geographers*, 95 (2), 411–36.

Miao, Ye. 2003. "China's land market creates pressures", *Asia Times*, 20 August.

Nagendra, Harini and Ostrom, Elinor 2008 "Governing the commons in the new millennium: A diversity of institutions for natural resource management", in, Cutler J. Cleveland (ed.), *Encyclopedia of Earth*, http://www.eoearth.org/article/Governing_the_commons_in_the_new_millennium:_A_diversity_of_institutions_for_natural_resource_management, accessed 12 November 2008.

Nanfang Daily. 2008. "珠三角最后的人民公社" (The last commune in the Pearl River Delta), 29 May, http://www.nanfangdaily.com.cn, accessed 11 August 2008.

Oi, Jean C. 1999. *Rural China Takes Off: Institutional Foundations of Economic Reform*, Berkeley, Los Angeles and London: University of California Press.

Oomen, Barbara. 2002. *Chiefs! Law, Power and Culture in Contemporary South Africa*, Leiden: Leiden University Press.

Ostrom, Elinor. 2005. *Understanding Institutional Diversity*, Princeton and Oxford: Princeton University Press.

Pei, Minxin. 2006. *China's Trapped Transition: The Limits of Developmental Autocracy*, Cambridge MA: Harvard University Press.

Perry, Elizabeth J. and Selden, Mark (eds). 2003. *Chinese Society: Change, Conflict and Resistance*, 2nd edition, London and New York: Routledge.

Prosterman, Roy, Hanstad, Tim and Li, Ping. 1998. "Large-scale farming in China: an appropriate policy?", *Journal of Contemporary Asia*, 28 (1), 74–102.

Scott, James C. 1985. *Weapons of the Weak: Everyday Forms of Peasant Resistance*, New Haven: Yale University Press.

Southern Weekend. 2011. "最后的人民公社即将终结" (The last people's commune is to end), http://www.infzm.com/content/62767, accessed 17 December 2011.

Ubink, Janine M. and Quan, Julian F. 2008. "How to combine tradition and modernity? Regulating customary land management in Ghana", *Land Use Policy*, 25, 198–213.

von Benda-Beckmann, Franz and Keebet. 2006. "How communal is communal and whose communal is it? Lessons from Minangkabau", in Keebet and Franz von Benda-Beckmann and Melanie Wiber (eds), *Changing Properties of Property*, New York and Oxford: Berghahn Books.

Walker, Kathy le M. 2008. "From covert to overt: Everyday peasant politics in China and the implications for transnational agrarian movement", *Journal of Agrarian Change*, 8 (2, 3), 462–88.

Wang, Zhaoping. 2007. 转型经济发展中的文化断裂与贫困研究 (*Cultural Rupture and Poverty Studies in Transitional Economy*), Beijing: China Social Science Press.

Yakou Village Administrative Committee. 2005. *Yakou Village Record* (internal publication and circulation).

Zhang, Xiaojun. 2004. "象征地权与文化经济: 福建杨村的历史地权个案研究" (Symbolic land rights and cultural economy: Case study on historic land rights in Yang Village, Fujian Province), *China Social Sciences*, 3, http://big5.xjass.com/jj/content/2008-08/01/content_26484.htm, accessed 15 December 2008.

Zoomers, Annelies and van der Haar, Gemma (eds). 2000. *Current Land Policy in Latin America: Regulating Land Tenure Under Neo-liberalism*, Amsterdam: Royal Tropical Institute.

Chapter 8

Conditions and Dynamics of Pro-poor Land Tenure

Land Tenure, Rural Development and Governance Linkages

Facing the challenges of rural development, and rampant farmland loss in particular, the Chinese government has never ceased its efforts to reform the country's land laws and policies. Despite the progress made in ensuring equal distribution of land rights and support for household-based farming to improve peasants' livelihoods and agricultural production, this book intends to show that there are no one-size-fits-all solutions to sustainable land resource use and management in general, and land tenure institutions in particular. In most cases where the peasants cannot utilize the land to its full potential, the absence of an enabling environment to ensure their participation in land use and planning processes further undermines the effectiveness of the land reform. The existing polices lack specific mechanisms that allow the peasants to effectively organize themselves in forging viable relations that suit their best interests. The current state-led and market-oriented land institutions have yet to encourage the establishment of peasant-centred arrangements for sustainable land utilization. Policy failure in tackling complex meanings of land, notions of property relations and the underlying social, political and economic contexts further sets structural limits to current reform measures (see Sikor and Müller, 2009).

This book provides a holistic approach to an understanding of evolving land rights, land institutional change and sustainable land use and rural development, especially focusing on the historical, political and social dimensions of the past and present practices of land management in China. It contributes to a multi-dimensional study of rural development and land reform linkages which are interpreted differently by different stakeholders. For the peasants in the vast poor rural areas, land remains their basic means of subsistence given a lack of social support programmes to provide a social safety net for them in the face of land expropriation. Yet this does not mean that they have the intention to stick to the land as it is. Their decisions over land use are contingent upon numerous external factors; and in most cases, their lack of alternative choices actually frees up ample space for the local state to manipulate the entire process of land use and management. The mere absence of peasant-initiated activities and organizations explains their vulnerability to any infringement of their rights by the local state and corporations. For the state, land acquisition and expropriation still constitute the most important means of local financing and reaching the goal of rapid local economic growth, while paying a high cost in terms of peasants' land tenure insecurity that triggers social unrest. In the process of decentralization, the local state continues to experience fiscal constraints on economic and social development, which lead to a quick fix through land sales to obtain lucrative revenues. The downward cycle of land loss further exacerbates rural poverty and social inequality between the rural and urban dwellers. In China where legal and policy instruments have not functioned well to the benefit of the poor, land tenure insecurity and poor rural governance continue to constrain China's path towards sustainable rural development.

Few studies have addressed the linkages between land tenure, rural development and governance in a broader sense in order to understand the conditions and dynamics of pro-poor land tenure. The central tenet of this book is that land tenure, no matter what forms it takes, can possibly be sustainable in a local setting where it fits specific development, governance and resource use conditions. These intermingled conditions determine the dynamics of land tenure systems. More pro-poor and dynamic land tenure systems can also contribute to the improvement of development, governance and resource use. To explicate this argument, this study takes a critical look at the historical implications of China's land reform, the trajectory of socialist-centred and market-oriented land laws and policies, local interpretations and implementation of the laws and policies, and local practices with elements of innovative choice over land use. It demonstrates that land property rights and land tenure security cannot be addressed adequately without the exploration of the institutions that support policy implementation and local innovation. And these institutions ought to be designed by the peasants themselves. Of course, the supporting roles of the state and businesses are needed. However, this remains a daunting challenge for them especially in view of the lack of legal and policy mechanisms. This is compounded by the lack of genuine village democratic governance that further hinders the creation of peasant incentives in the land reform process. In the Chinese social and political context, the market-oriented land reform, if not governed properly by the state, can also exacerbate land loss of the poor whose capacity in organizing themselves towards better land management is likely to be further undermined. There is urgency for the policy-makers to revisit their policies and examine more critically what the peasants need from land reform and how they can be empowered to participate in the process. Otherwise, in the name of improving land governance, the local state can find ways to take the laws and policies into their own hands and create those 'fancy' institutions that can do harm to the livelihoods of the poor.

Overarching Issues of Concern

To understand China's land reform, one needs to start from its history and the implications for current land institutional development. This book demonstrates that *social inequality* derived from land tenure has marked the struggles between the Chinese peasantry and the state. Essentially, these struggles underscore the partial failure of the land reform in uplifting peasants' social, political and economic status. Being economically poor without entitlements to land, the peasants are marginalized in the mainstream rural economic development. Their vulnerability to land loss and natural and economic shocks further exacerbates their poverty. In most cases, their poverty has advantaged the rural elite and local businesses, which rely on peasants' land for speculative gains. Land reform in the Ming and Qing dynasties spawned an alliance between the empire and local landlordism that put the poor landless peasants and tenants on the margin of development. Although various land reform measures were undertaken to curtail the power of the local state and landlordism, the peasantry was unable to forge a consistent strong force against their masters except for ad hoc cases of victories.

By contrast, the communist-led land revolution marked the beginning of a new era of China's social and political transformation characterized by 'land for the tiller' programmes. In the aftermath of the 1949 communist victory, land was equally redistributed among the peasant households. Strikingly, this move was aborted just a few years later by the introduction of the commune system. This means that the system of 'land for the tiller' had not brought about significant rural economic changes. Moreover, the commune system also ended in the failure to reorganize the peasants to achieve better agricultural production outcomes. It was then replaced by market-oriented reforms in which the Household Responsibility System (HRS) was introduced which aimed at creating peasant incentives in farming. Again, the HRS has not proven to be an effective solution to the complex rural problems in China. In short, all these land reform measures have one thing in common, that is, reform imposed from the top without effectively addressing the needs of the poor and tackling the *social and political structural factors constraining rural development*. Chinese rural societal organization has never been enabled to decide on the desired forms of land reform. Land reform imposed by the state has actually served the state's political needs more in terms of reorganization of the rural masses to consolidate control rather than facilitate peasant-centred pro-poor land institutional changes.

The ultimate goal of land governance ought to ensure poverty alleviation and sustainable rural development for the poor irrespective of the type of land reforms implemented. In this respect, no one-size-fits-all solutions to China's complex rural problems can be found. Rather, peasant institutional innovation ought to be encouraged and fostered. But first, the political will to do so is not clear; nor is it easy for the local state to implement because of different incentives to govern the land. Second, there are few demonstration pilots available to influence policy. Third, it remains a challenge to implement multi-stakeholder participation in land use and management. Last, how to work with the peasants and how the peasants can organize themselves to use and manage their land remains unaddressed in policy. The ongoing debates on landownership especially the focus on the clarification of collective rights and ownership as well as radical calls for privatization present a *simplistic approach to institutional reform*. This approach ignores the fundamental constraints to rural governance which is still characterized as top-down, although village democratization has brought about limited advances in peasant empowerment. Moreover, to a certain extent, the current market-oriented land laws and policies have actually co-existed with the fragmented social and political relations among the peasants whose collective choice and power over land use and management are dramatically undermined. Thus, the incentives and social and political realities of peasants ought to be taken into more serious consideration in any policy changes.

Local Practices and Experimentation

China does not lack local policy practices and experimentation, but does lack those more people-centred initiatives to provide more viable solutions. This book shows that more appropriately designed land tenure systems hinge on their compatibility with local conditions; otherwise, they are not sustainable or may even cause unintended consequences. The daunting challenges of sustainable land use and rural development facing the poverty regions of China put into question more individualized land tenure systems that provide limited mechanisms for collective action in poverty reduction and sustainable natural resource management. The individualization of tenure, at least in the Chinese context, does not fully reflect the need to shift to commercial agriculture, as most farmland acquisitions are for non-agricultural purposes (see Chimhowu and Woodhouse, 2006). It appears to reinforce individual interests and incentives rather than facilitate participatory institutions for consensus-building among the key stakeholders for the sake of sustainable rural development. First, stakeholders' conflicting interests show their different views on how land ought to be sustainably managed. As a result, the HRS has failed to stimulate the stakeholders to reach common objectives and strategies in land management. Second, the HRS is found to be an inappropriate approach to land resources management especially concerning rangeland, forestry and agricultural land use, as the management of these resources relies on more integrated institutions and collective choice for sustainable solutions. Third, it has further fragmented social and political relations characterized by rising relational rifts among individual peasant households, which make multi-stakeholder collaboration in land management more difficult. In short, the HRS is not a panacea to China's complex land problems, as its underlying social fragmentation facilitates poor land and village governance and undermines strong mechanisms of collective action. Thus, policy improvements should take account of the existing institutions and practices on the ground (see Sikor and Müller, 2009).

The strengthening of the HRS is seen as a further step towards local experimentation on the institution of land shareholding cooperatives. The latter, however, may have served the needs of the local state in land expropriation more than the so-called scaled development that would benefit the poor. As land shareholding cooperatives require reorganization of peasants' land rights and agricultural production, the role of the local state in carrying out this institutional change becomes paramount because of many contentious issues of poor governance that marginalize the majority of the poor peasant shareholders. The current practices are far from a peasant-centred collective treatment; yet they provoke a rethinking of inter-related issues of land policy and institutions, poverty and village governance. Moreover, the tendency towards this institutional development does show the weakening of the HRS in tackling land fragmentation and its underlying social and economic issues. Rather than strengthening the collective rights of the peasants, land shareholding cooperatives may reinforce the power of the local state in the absence of participative village governance processes to hold the state to account.

By contrast, the more grassroots level of institutional innovation in the case of the Yakou commune illustrates peasant-centred land tenure alternatives to that of the earlier mentioned technocratic and bureaucratic approaches. It underscores the value of peasant self-help in organizing and representing the community as an effective collective force against local state interference in land management. Standing in opposition to the mainstream economy, the economic system of the commune based on egalitarian principles and practices further ensures equal distribution of village assets and wealth derived from land use among the commune members. Accountable and transparent village governance plays an essential role in social, political and economic processes concerning land use. In particular, the strong village leadership provides the commune members with ample space to participate in the governance process. It demonstrates that community-centred collective action still provides a viable alternative in conjunction with market-oriented economic reform, whilst making use of the advantages that the latter offers.

However the recent village developments, especially with regard to the resignation of the village Party Secretary, calls into question the perpetuation of the commune, which reflects the mounting challenges of village economic development and the conflicting interests in land use and power struggles among the peasants and their leaders and the local state (see Agrawal and Gibson, 1999; Leach et al., 1999). Moreover, this issue evokes rethinking of the meaning of land tenure, development and governance for these actors and the overall development trajectory of the state. Any changes in the policy and law of the latter will continue to challenge the commune that still needs to operate more effectively to more adequately cope with the mainstream political economy of the region. Despite these challenges, the existence of the commune to date is a critical case in point for understanding the dynamics and conditions of locally-based land tenure, which is essential to tackle the structural constraints introduced by the ongoing market-oriented institutional reform in rural China.

In making use of empirical cases this book intends to strongly argue that although land tenure is a social and political relationship, or is embedded in these relations, it is more important to realize how it is reshaped by economic and political forces in face of rural and urban development. It is in these complex relations that an individualized tenure may do more harm than good to the land rights and livelihoods of the peasants. Neither would state-formulated experimentation on land tenure arrangements offer the needed space for people-centred institutional innovation. The approach to exploring the land tenure, rural development and governance linkages in terms of the conditions and dynamics of pro-poor land tenure is intended to offer rethinking of the current debates about China and even other developing countries concerning the relationships between land reform, agrarian transformation and social equity (see Peters, 2009: 1318).

Key Research Issues Revisited

In essence, a particular land tenure regime can only be sustained provided that it complies with the needs of the peasants for sustainable rural development in different local contexts, which demand for more rural institutional innovation. To avoid generating partial understanding of land-related rural problems through a single disciplinary lens, this book seeks to develop a relatively comprehensive scenario of land policy and practice while maintaining its anthropological and political economic focuses. It pays attention to four critical issues in China's land reform process, namely: responsiveness to local livelihoods; connections with dynamics of authority; interactions with social inequalities; and environmental repercussions (see Sikor and Müller, 2009: 1312). How to make land tenure work for the poor in the name of rural sustainable development is the ultimate challenge for the Chinese policy-makers. Thus, the findings of this study further address the key issues pertinent to the cases of other countries, and transition economies in particular.

Whose Common Property?

Common property often refers to property as jointly owned and managed by groups. Common property regimes are criticized for their contribution to land tenure insecurity and to obstructing economic development. Instead, states and markets are seen as the appropriate institutional avenues to address policy failures in natural resources management (see Shapiro, 1989).

China's collective landownership carries the characteristics of common property to a certain extent. Although rural farmland is contracted out to individual households, it is still under the overall management responsibility of the village collective. The political, social and economic relationships between the individual household land users and the collective become complicated and evolve over time and space. However, more individually-oriented land policies may not be appropriate given regional diversities in economic and social development. Furthermore, the issue of the 'tragedy of the commons' is inappropriately applied across the board (Hardin, 1968). As such, collective landownership has its particular relevance for Chinese politics and society. The reality of a large population living in poverty and with limited land and other natural resources requires interactions between the state and the local community, and collective action towards sustainable land utilization and livelihoods. The mixed pattern of collective landownership with individual peasant's land use rights has enabled the state to formulate land policies to cater for its social and economic development needs. Despite their induced problems, at least the collective is still legally recognized and plays an essential role in organizing the peasant society. However, the challenge remains as to how to make it work more effectively for the individual poor households and to stimulate more meaningful collective institutional arrangements for land use and management, as the collective has represented the interests of the state more than that of the poor. This requires more in-depth studies of changing property rights relations as part of the broader social, political and economic changes (Shipton and Goheen, 1992; Ensminger, 1997; Hann, 1998).

The existence of common property regimes in many parts of the world reflects the importance of social relations as complex dimensions of land tenure. Social relations in the Chinese context exhibit an interesting area for the study of the role of land in reconstructing the relations among various stakeholders. The Chinese countryside reveals the existence of both fragmented land relations and the predominance of rural collectives and village administrative representative committees. Linking these dimensions to land property rights, one finds that the current policy focused on strengthening individual's land rights may further exacerbate land fragmentation and loosen community coherence. This may favour the powerful actors who impose unfavourable conditions on the poor, since the latter find it difficult to organize collective action. Ultimately, if the law does not provide ample impetus for community-organized land relations and collective action, the trend towards poor land and village governance cannot be averted. This study tries to address the underlying challenges for land institutional innovation constrained by the asymmetric power relations between the peasants and other stronger stakeholders. This is in line with the latest developments in common property studies (see Varughese and Ostrom, 2001; Agrawal, 2005).

Common property regimes should not be assessed only from negative perspectives. As Ostrom (1990) argues, the study of common property regimes reveals the micro-institutional regulation of resources and the possibilities of community, especially small groups of resource users who are able to craft viable forms of resources governance. This is exactly the case of Yakou commune which shows that the institution of private land property rights should not be treated as a teleological and deterministic logic to China's agrarian future. This is because concepts such as private, public or common are just too general to sufficiently reflect the variations in local institutional resource governance (McKean, 1992; Agrawal, 2005). Future research on the public/collective and private land tenure interfaces and their linkages with the broader issues of the political economy and governance and social processes would contribute to land policy improvements for the Chinese peasantry. This means that there should not be any prescriptions of individual or collective tenure for a given setting; neither would it be simply a matter of the choice of local community given the intra- and inter-group conflicts and appropriation of land by representatives of the state (see Platteau, 2000). Ultimately, a multi-stakeholder approach to reforming land tenure in tackling the broader challenges of rural and even urban development that prioritize the needs of the rural communities for enhanced and sustainable livelihoods is needed.

Bundles of Rights versus Bundles of Power

Land tenure security is often recognized as the fundamental issue to be tackled for successful land reform programmes which would cultivate more rights for the poor. Thus, the concept of bundles of rights has received wide recognition in the study of people–land relations. However, this study shows that no matter how strong those rights are enshrined in laws, when power is not given to the individual peasant households and collectives, their rights can easily be abused by the powerful state, corporations and local elites. Thus, a bundle of power instead of the property notion of a bundle of rights is more useful for poor peasants (see Ribot and Peluso, 2003). This finding has important implications for any pro-poor land reforms which ought to empower the poor both as individuals and as a group to have a stronger voice in the land reform process.

As a corollary, the practice of land registration aimed at the demarcation of land boundaries and clarification of an individual's land rights has limitations in addressing the wider complex social and political relations between the landowners or users and other actors. That is why in many countries land registration projects have failed to protect the rights of the poor and they have reinforced the existing inequality between different groups (McAuslan, 2003). Without first tackling the power imbalances among different groups, this approach will be ineffective in addressing the fundamental issues of economic and social inequality within a given community.

Following the 2007 Property Law, land registration in China has been executed at the village collective level, for the rural land is collectively owned. As such, the power of the village administrative committee as the registrant vis-à-vis the peasant households is strengthened. Simply, this reflects the fact that the state's interest and power in strengthening individual peasants' land rights is not sufficiently reflected in the practice of land registration. Rather, land registration serves the purpose of land administration for technical purposes. Its drawback lies in its static and technocratic approach which excludes the flexibility to address household needs. However, it is an easy approach to land administration for the state, since any complications in land registration may unavoidably touch upon the complex issue of bundle of rights and power underpinning land registration. Furthermore, there is a certain level of mistrust between the local state, the village administrative committee and the peasants. Further research on the dynamics of rural power relations that shape property relations among diverse social actors will shed more light on the structural constraints on legal empowerment of the poor in China in land governance processes.

Land and Social Capital

China's social, economic and environmental problems cannot be tackled with simplistic approaches. Rather, the use of comprehensive and systemic approaches should be explored. Poverty alleviation still counts as a daunting task for the government. While promoting market-oriented land reform policies, the government has acted with due caution. However, it has not managed to put forward more viable solutions. This study shows that there is a need to further understand the particularity of each village and region in terms of the livelihood patterns and the impact factors. This would require a greater level of flexibility in allowing for local experimentation in farmland use and management. More properly designed land shareholder cooperatives can be seen as a major mechanism for improved land management and utilization for the benefit of all. In this regard, individual land rights can be coupled with group rights, which will allow for more voluntary organized peasant cooperatives. Certainly, this remains an unaddressed political issue not only in China but also in many other developing countries.

Furthermore, this study reveals the fact that peasants' lack of voice in the political system exacerbates their land tenure insecurity. Current land reform measures, which do not address the key power asymmetries between the peasantry and the state, can only undermine the incentives of peasants in participating in village governance and development processes. When these challenges remain untracked, any attempt to address land tenure security would be of little practicality. Moreover, land is not solely related to poverty, sustainable livelihoods and natural resources management, especially when peasants lack adequate access to capital and public services for agricultural production and social welfare. In this situation, social capital becomes paramount in enabling the peasants to organize themselves in combating various natural, economic and political constraints on poverty alleviation in dealing with inadequately developed markets and other issues (Amarasinghe, 2009). As genuine self-organized peasant organizations are lacking in China, the predominance of unequal economic and political relations between the peasants and the state will continue to hinder more pro-poor land institutional development.

Further Policy Implications

The challenges for China's land reform and sustainable rural development call for significant rethinking of land governance mechanisms. In particular, the failure of the current legal environment to recognize these challenges and put forward more diverse approaches reflects the weakness of the Chinese society in participating in policy and law-making processes. To improve this situation, it is important that the law should not only serve the need for economic reform and the interests of the policy-makers, but also address the fundamental barriers to social mobilization and individual and collective power. Moreover, it ought to further promote, encourage and stimulate peasant self-initiated activities in the use of land to meet their basic needs. This would require more interactions between the law-makers and the people in the law-making process. Also the law ought to create a viable framework that does not restrict the prevailing livelihood practices. As seen in many parts of the world, the asymmetry between law and the lived experiences of the rural poor can lead to unintended consequences at the expense of the poor (see Agrawal, 2005).

Land policy is ultimately about society and the organization and governance of relationships between people (see McAuslan, 2003). Any attempt to strengthen land governance without paying attention to the local relational context would result in inefficient, costly and possibly adverse consequences. As the Chinese government is gaining ground in executing the so-called world's strictest land management policies to tackle land mismanagement and farmland loss, more appropriate policy measures based on local practices are needed. This would require a strong commitment of the state to support society at large (see Tessemaker and Hilhorst, 2007). A pro-poor land policy to recognize and balance the diverse interests of different stakeholders to avoid favouring one particular group while disadvantaging the other is urgently needed. To this end, the state has to confront the very structures that perpetuate the existing unfavourable conditions for the poor (Borras and Franco, 2010: 11, 23).

Land policy-making processes should support more meaningful public debates with the peasants and other stakeholders, exchange of experiences and pilot testing of innovative approaches. The latter would allow for the state's facilitation of considerable flexibility for the local community to manoeuvre their use of land and cope with uncertainty around land tenure, land use and management (see Meinzen-Dick and Pradhan, 2001). Given China's rural reality, it is important to find ways to revitalize the rural economy characterized by smallholder production while addressing social inequality and differentiation. This is a key challenge for the state and Chinese society to work together to put forward an agenda for action. Strong commitment by the state and public investment are necessary. But without a fully-fledged civil society and pro-poor land institutions, land tenure insecurity and vulnerability to poverty and natural resource degradation and depletion are envisaged to persist for a long time for the Chinese peasantry.

References

Agrawal, Arun. 2005. *Environmentality: Technologies of Government and the Making of Subjects*, Durham and London: Duke University Press.
Agrawal, Arun and Gibson, Clark. 1999. "Enchantment and disenchantment: The role of community in natural resource conservation", *World Development*, 27, 629–49.

Amarasinghe, Oscar. 2009. "Social capital to alleviate poverty: Fisheries cooperatives in southern Sri Lanka", *The Newsletter*, 51 (15), International Institute for Asian Studies,

Borras Saturnino M. and Franco, Jennifer C. 2010. "Contemporary discourses and contestations around pro-poor land policies and land governance", *Journal of Agrarian Change*, 10 (1), 1–32.

Chimhowu, Admos and Woodhouse, Phil. 2006. "Customary vs. private property rights? Dynamics and trajectories of vernacular land markets in sub-Saharan Africa", *Journal of Agrarian Change*, 6 (3), 346–71.

Ensminger, Jean. 1997. "Changing property rights: Reconciling formal and informal rights to land in Africa", in J.N. Drobak and J.V.C. Nye (eds), *The Frontiers of the New Institutional Economics*, San Diego: Academic Press.

Hann, Chris (ed.). 1998. *Property Relations: Renewing the Anthropological Tradition*, Cambridge: Cambridge University Press.

Hardin, Garret. 1968. "The tragedy of the commons", *Science*, 162, 1243–8.

Leach, Melissa, Mearns, Robin and Scoones, Ian. 1999. "Environmental entitlements: Dynamics and institutions in community-based natural resource management", *World Development*, 27 (2), 225–47.

McAuslan, Patrick. 2003. *Bringing the Law Back in: Essays in Land, Law and Development*, Aldershot: Ashgate.

McKean, Margaret. 1992. "Success on the commons: A comparative examination of institutions for common property resource management", *Journal of Theoretical Politics*, 4 (3), 247–81.

Meinzen-Dick, Ruth S. and Prahan, Rajendra. 2001. "Implications of legal pluralism for natural resource management", in Lyla Mehta, Melissa Leach and Ian Scoones (eds), *Environmental Governance in an Uncertain World, IDS Bulletin*, 32 (4), 10–17.

Ostrom, Elinor. 1990. *Governing the Commons: The Evolution of Institutions for Collective Action*, New York: Cambridge University Press.

Peters, Pauline E. 2009. "Challenges in land tenure and land reform in Africa: anthropological contributions", *World Development*, 37 (8), 1317–25.

Platteau, Jean-Philippe. 2000. "Does Africa need land reform?", in C. Toulmin and J. Quan (eds), *Evolving Land Rights, Policy and Tenure in Africa*, London: Department for International Development (DFID)/IIED/NRI.

Ribot, Jesse and Peluso, Nancy Lee. 2003. "A theory of access", *Rural Sociology*, 68 (2), 153–81.

Shapiro, Ian. 1989. "Gross concepts in political argument", *Political Theory*, 17 (1), 51–76.

Shipton, Parker and Goheen, Mitzi. 1992. "Understanding African land-holding: Power, wealth and meaning", *Africa*, 62 (3), 307–25.

Sikor, Thomas and Müller, Daniel. 2009. "The limits to state-led land reform: An introduction", *World Development*, 37 (8), 1307–16.

Tessemaker, Esther and Hilhorst, Thea. 2007. *Netherlands' Support to Strengthening Land Tenure Security in Developing Countries: Overview and Lessons Learned*, Den Haag: Ministry of Foreign Affairs of the Netherlands.

Verughese, George and Ostrom, Elinor. 2001. "The contested role of heterogeneity in collective action: Some evidence from community forestry in Nepal", *World Development*, 29 (5), 747–65.

Index